Engineering Design Methods

Strategies for Product Design

FOURTH EDITION

Engineering Design Methods

Strategies for Product Design

FOURTH EDITION

Nigel Cross
Open University, UK

John Wiley & Sons, Ltd

Other Wiley Editorial Offices

Library of Congress Cataloging in Publication Data

Cross, Nigel, 1942–
    Engineering design methods : strategies for product design/Nigel Cross.—4th ed.
        p.   cm.
    Includes bibliographical references and index.
    ISBN 978-0-470-51926-4 (pbk. : alk.paper)
    1. Engineering design.   I. Title.
    TA174.C76 2008
    620′.0042—dc22

                                                            2008002727

British Library Cataloguing in Publication Data

A catalogue record for this book is available from the British Library

ISBN 978-0-470-51926-4 (P/B)

Typeset in 10/13.5pt Palatino by Integra Software Services Pvt. Ltd, Pondicherry, India
Printed and bound by CPI Group (UK) Ltd, Croydon, CR0 4YY

Contents

About the Author

Professor Nigel Cross is a leading international figure in the world of design research and methodology. He is a long-time member of the academic staff of the UK's pioneering, multimedia Open University, where he has been involved in developing a wide range of distance-education courses in design. He has also held visiting appointments in The Netherlands, Australia and the USA. His research interests are principally in understanding the nature of design ability, and the development of design skill from novice to expert. Professor Cross is also editor-in-chief of *Design Studies*, the international journal of design research.

Acknowledgements

The author and publisher gratefully acknowledge the following sources and permissions to reproduce figures:

Figure 1.1: T.A. Thomas, *Technical Illustration*, McGraw-Hill. Figure 1.2: J. Fenton, *Vehicle Body Layout and Analysis*, Mechanical Engineering Publications. Figure 1.3: A.L.H. Dawson. Figure 1.4: A. Howarth. Figure 1.6: S. Dwarakanath and L. Blessing, in *Analysing Design Activity*, Wiley. Figure 2.1: Biblioteca Ambrosiana, Milan. Figure 2.2: C. Moore and Van Nostrand Reinhold. Figure 3.2: M.J. French, *Conceptual Design for Engineers*, Design Council. Figures 3.3, 3.4, 3.5: B. Hawkes and R. Abinett, *The Engineering Design Process*, Longman. Figures 3.8, 16.4, 7.5, 7.6, 7.8, 8.1, 8.2, 10.1, 11.3, 11.4, 11.5: G. Pahl and W. Beitz, *Engineering Design*, Design Council/Springer-Verlag. Figures 3.9, 3.10: VDI-Verlag. Figure 3.11: L. March, *The Architecture of Form*, Cambridge University Press. Figure 5.1: Kenneth Grange. Figure 5.2: IDEO. Figure 6.2: J.C. Jones, *Design Methods*, Wiley. Figure 16.5: E. Tjalve, *A Short Course in Industrial Design*, Butterworth. Figure 16.6: S. Pugh, *Total Design*, Addison Wesley. Figures 7.3, 7.4: E. Krick, *An Introduction to Engineering*, Wiley. Figure 7.7: K. Otto and K. Wood, *Product Design*, Pearson Education. Figure 8.3: S. Love, *Planning and Creating Successful Engineered Designs*, Advanced Professional Development. Figure 8.4: K. Hurst, *Engineering Design Principles*, Edward Arnold. Figures 9.1, 11.9: D.G. Ullman, *The Mechanical Design Process*, McGraw-Hill. Figures 9.2, 9.3: R. Ramaswamy and K. Ulrich, in *Design Theory and Methodology 92*, ASME. Figures 9.4, 9.5: J.R. Hauser and D. Clausing, *Harvard Business Review*. Figures 10.2, 10.3: V. Hubka, *Principles of Engineering Design*, Butterworth. Figures 10.4, 10.5: K. Ehrlenspiel, in *ICED 87*, Heurista. Figures 10.6, 10.7: M. Tovey and S. Woodward, *Design Studies*, Elsevier Science. Figure 11.2: U. Pighini, *Design Studies*, Elsevier Science. Figures 11.6, 11.7, 11.8: M. Axeborn, A. Gould and S. Ottoson, *Journal of Engineering Design*, Taylor & Francis, informaworld.com. Figures 11.10, 11.11, 11.12: K. Ulrich and

S. Eppinger, *Product Design and Development*, McGraw-Hill. Figures 12.1, 12.2: T.C. Fowler, *Value Analysis in Design*, Van Nostrand Reinhold. Figures 12.3, 12.4: Engineering Industry Training Board. Figure 12.5: A.H. Redford, *Design Studies*, Elsevier Science. Figures 12.6, 12.7, 12.8: Open University. Figure 14.3: British Standards Institution. Figure 14.4: Richard Hearne.

Introduction

This book offers a strategic approach and a number of tactics as aids for designing successful products. It is intended primarily for use by students and teachers of engineering design and industrial design. Its main emphasis is on the design of products that have an engineering content, although most of the principles and approaches that it teaches are relevant to the design of all kinds of products. It is essentially concerned with problem formulation and the conceptual and embodiment stages of design, rather than the detail design which is the concern of most engineering texts. The book can most effectively be used in conjunction with projects and exercises that require the formulation and clarification of design problems and the generation and evaluation of design solutions.

This fourth edition of the book is again an improved, revised and updated version. A new design method, the user scenarios method, has been added in the central part of the book. The inclusion of this method completes the circle of the integrative model of the design process around which the method chapters are structured. Elsewhere in the methods chapters (Chapters 5–12), new examples of the application of design methods in practice have also been introduced. The book is much more than a manual of procedures: throughout there is discussion and explication of the principles and practice of design.

The contents of the book are divided into three parts. Part One, Understanding Design, provides an overview of the nature of design activity, designers' natural skills and abilities, and models of the design process. Chapter 1 introduces the kinds of activities that designers normally undertake, and discusses the particular nature and structure of design problems. Chapter 2 considers and discusses the cognitive abilities that designers call upon in tackling design problems, and outlines some of the issues involved in learning and developing these 'designerly' skills and abilities. Chapter 3 reviews several of the models of the design process that have been developed in order to help designers structure their approach to designing, and suggests a new hybrid, integrative model that combines both the procedural and the structural aspects of the nature of design.

Part Two, Doing Design, explains the details of how to do design, at various stages of the design process. Chapter 4 reviews the field of design methods, describes a number of methods that help to stimulate creative design thinking and introduces the rational methods that are presented in the following chapters. Chapters 5–12 constitute a manual of design methods (the tactics of design), presented in an independent-learning format, i.e. students can be expected to learn the principal features of the methods directly from the book. These eight chapters follow a typical procedural sequence for the design process, providing instruction in the use of appropriate methods within this procedure. Each chapter presents a separate method, in a standard format of a step-by-step procedure, a summary of the steps and a mixed set of practical examples concluding with a fully worked example. The eight methods included are: user scenarios; objectives tree; function analysis; performance specification; quality function deployment; morphological chart; weighted objectives; and value engineering.

Part Three, Managing Design, is concerned with managing the design process, from both the product designer's viewpoint and the business manager's. Chapter 13 outlines a strategic approach to the design process, utilizing the most appropriate combination of creative and rational methods to suit the designer and the design project. Reflecting the approach that is implicit throughout the book, the emphasis is on a flexible design response to problems and on ensuring a successful outcome in terms of good product design. Chapter 14 puts the role of design into a broader perspective of new product development, showing that successful product design is framed on the one hand by business strategy and on the other hand by consumer choice.

The book embodies a concept of 'product design' that combines the two more traditional fields of engineering design and industrial design: the new concept of 'industrial design engineering'. Although intended primarily for students of product design – no matter whether their courses are biased more towards engineering or industrial design – the book is also useful as an introduction to design for the many teachers and practitioners in engineering who found this subject sadly lacking in their own education.

Part One
Understanding Design

1 *The Nature of Design*

Design Activities

People have always designed things. One of the most basic characteristics of human beings is that they make a wide range of tools and other artefacts to suit their own purposes. As those purposes change, and as people reflect on the currently available artefacts, so refinements are made to the artefacts, and sometimes completely new kinds of artefacts are conceived and made. The world is therefore full of tools, utensils, machines, buildings, furniture, clothes and many other things that human beings apparently need or want in order to make their lives better. Everything around us that is not a simple, untouched piece of nature has been designed by someone.

In traditional, craft-based societies the conception, or 'designing' of artefacts is not really separate from making them; that is to say, there is usually no prior activity of drawing or modelling before the activity of making the artefact. For example, a potter will make a pot by working directly with the clay, and without first making any sketches or drawings of the pot. In modern, industrial societies, however, the activities of designing and of making artefacts are usually quite separate. The process of making something cannot normally start before the process of designing it is complete. In some cases, for example in the electronics industry, the period of designing can take many months, whereas the average period of making each individual artefact might be measured only in hours or minutes.

Perhaps a way towards understanding this modern design activity is to begin at the end; to work backwards from the point

Engineering Design Methods: Strategies for Product Design   N. Cross
© 2008 John Wiley & Sons, Ltd

where designing is finished and making can start. If making cannot start before designing is finished, then at least it is clear what the design process has to achieve. It has to provide a description of the artefact that is to be made. In this design description, almost nothing is left to the discretion of those involved in the process of making the artefact; it is specified down to the most detailed dimensions, to the kinds of surface finishes, to the materials, their colours, and so on.

In a sense, perhaps it does not matter how the designer works, so long as he or she produces that final description of the proposed artefact. When a client asks a designer for 'a design', that is what they want: the description. The focus of all design activities is that endpoint.

Communication of designs

The most essential design activity, therefore, is the production of a final description of the artefact. This has to be in a form that is understandable to those who will make the artefact. For this reason, the most widely used form of communication is the drawing. For a simple artefact, such as a door-handle, one drawing would probably be enough, but for a larger, more complicated artefact such as a whole building the number of drawings may well run into hundreds, and for the most complex artefacts, such as chemical process plants, aeroplanes or major bridges, then thousands of drawings may be necessary.

These drawings will range from rather general descriptions, such as plans, elevations and general arrangement drawings, that give an 'overview' of the artefact, to the most specific, such as sections and details, that give precise instructions on how the artefact is to be made. Because they have to communicate precise instructions, with minimal likelihood of misunderstanding, all the drawings are themselves subject to agreed rules, codes and conventions. These codes cover aspects such as how to lay out on one drawing the different views of an artefact relative to each other, how to indicate different kinds of materials and how to specify dimensions. Learning to read and to make these drawings is an important part of design education.

The drawings will often contain annotations of additional information. Dimensions are one such kind of annotation. Written instructions may also be added to the drawings, such as notes on the materials to be used (as in Figure 1.1).

Other kinds of specifications as well as drawings may also be required. For example, the designer is often required to produce

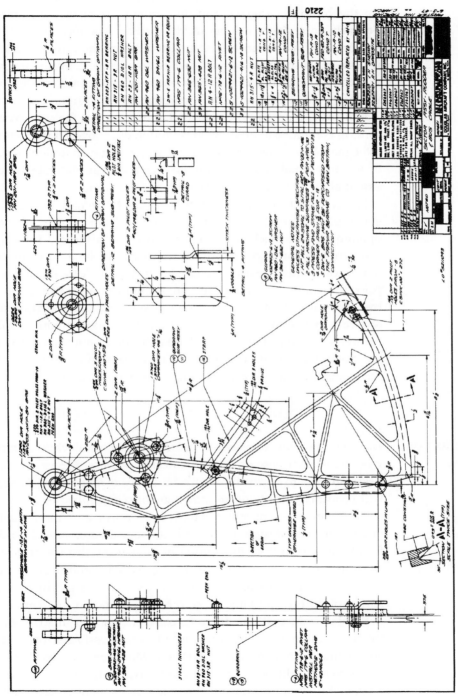

Figure 1.1 Communication: an example of a classic engineering design detail drawing

lists of all the separate components and parts that will make up the complete artefact, and an accurate count of the numbers of each component to be used. Written specifications of the standards of workmanship or quality of manufacture may also be necessary. Sometimes, an artefact is so complex, or so unusual, that the designer makes a complete, three-dimensional mock-up or prototype version in order to communicate the design.

However, there is no doubt that drawings are the most useful form of communication of the description of an artefact that has yet to be made. Drawings are very good at conveying an understanding of what the final artefact has to be like, and that understanding is essential to the person who has to make the artefact.

Nowadays it is not always a person who makes the artefact; some artefacts are made by machines that have no direct human operator. These machines might be fairly sophisticated robots, or just simpler, numerically controlled tools such as lathes or milling machines. In these cases, therefore, the final specification of a design prior to manufacture might not be in the form of drawings but in the form of a string of digits stored on a disk, or in computer software that controls the machine's actions. It is therefore possible to have a design process in which no final communication drawings are made, but the ultimate purpose of the design process remains: the communication of proposals for a new artefact.

Evaluation of designs

However, for the foreseeable future, drawings of various kinds will still be used elsewhere in the design process. Even if the final description is to be in the form of a string of digits, the designer will probably want to make drawings for other purposes.

One of the most important of these other purposes is the checking, or evaluating, of design proposals before deciding on a final version for manufacture. The whole point of having the process of design separated from the process of making is that proposals for new artefacts can be checked before they are put into production. At its simplest, the checking procedure might merely be concerned with, say, ensuring that different components will fit together in the final design; this is an attempt to foresee possible errors and to ensure that the final design is workable. More complicated checking procedures might be concerned with, say, analysing the forces in a proposed design to ensure that each component is designed to withstand the loads on it (Figure 1.2); this involves a process of

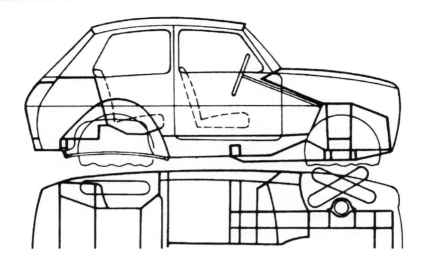

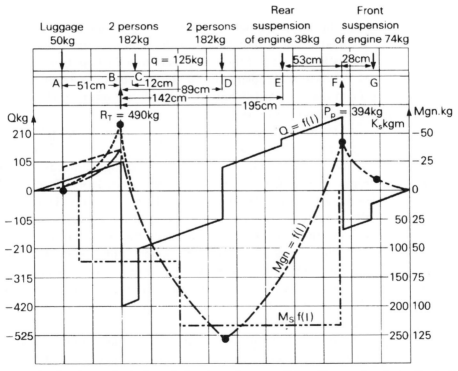

Figure 1.2 Evaluation: calculation of the shear forces and bending moments in the body of a small automobile

refining a design to meet certain criteria such as maximum strength or minimum weight or cost.

This process of refinement can be very complicated and can be the most time-consuming part of the design process. Imagine, for example, the design of a bridge. The designer must first propose the form of the bridge and the materials of which it will be made. In order to check that the bridge is going to be strong enough and stiff enough for the loads that it will carry, the designer must analyse the structure to determine the ways in which loads will be carried by it, what those loads will be in each member of the structure, what deflections will occur, and so on. After a first analysis, the designer might realize, or at least suspect, that changing the locations or angles of some members in the bridge will provide a more efficient distribution of loadings throughout the whole structure. But these changes will mean that the whole structure will have to be reanalysed and the loads recalculated.

In this kind of situation, it can be easy for the designer to become trapped in an iterative loop of decision-making, where improvements in one part of the design lead to adjustments in another part which lead to problems in yet another part. These problems may mean that the earlier 'improvement' is not feasible. This *iteration* is a common feature of designing.

Nevertheless, despite these potential frustrations, this process of refinement is a key part of designing. It consists, firstly, of analysing a proposed design, and for this the designer needs to apply a range of engineering science or other knowledge. In many cases, specialists with more expert knowledge are called in to carry out these analyses. Secondly, the results of the analysis are evaluated against the design constraints: does the design come within the cost limit, does it have enough space within it, does it meet the minimum strength requirements, does it use too much fuel?, and so on. In some cases, such constraints are set by government regulations, or by industry standards; others are set by the client or customer.

Many of the analyses are numerical calculations, and therefore again it is possible that drawings might not be necessary. However, specialists who are called in to analyse certain aspects of the design will almost certainly want a drawing, or other model of the design, before they can start work. Visualizations of the proposed design may also be important for the client and designer to evaluate aspects such as appearance, form and colour.

Generation of designs

Before any of these analyses and evaluations can be carried out the designer must, of course, first generate a design proposal. This is often regarded as the mysterious, creative part of designing; the client makes what might well be a very brief statement of requirements, and the designer responds (after a suitable period of time) with a design proposal, as if conjured from nowhere. In reality, the process is less 'magical' than it appears.

In most cases, for instance, the designer is asked to design something similar to that which he or she has designed before, and therefore there is a stock of previous design ideas on which to draw. In some cases, only minor modifications are required to a previous design.

Nevertheless, there is something mysterious about the human ability to propose a design for a new, or even just a modified, artefact. It is perhaps as mysterious as the human ability to speak a new sentence, whether it is completely new, or just a modification of one heard, read or spoken before.

This ability to design depends partly on being able to visualize something internally, in the 'mind's eye', but perhaps it depends even more on being able to make external visualizations. Once again, drawings are a key feature of the design process. At this early stage of the process, the drawings that the designer makes are not usually meant to be communications to anyone else. Essentially, they are communications with oneself, a kind of thinking aloud. As the example of the concept sketch for the 1950s Mini car shows (Figure 1.3), at this stage the designer is thinking about many aspects together, such as materials, components, structure and construction, as well as the overall form, shapes and functions.

Exploration of designs

At the start of the design process, the designer is usually faced with a very poorly defined problem; yet he or she has to come up with a well-defined solution. If one thinks of the problem as a territory, then it is largely unexplored and unmapped, and perhaps imaginary in places! As Jones (1992) has suggested, and as will be discussed in Chapter 13, it is therefore appropriate to think of the designer as an explorer, searching for the undiscovered 'treasure' of a satisfactory solution concept.

Equally, if one thinks of all potential solutions as occupying a kind of solution space, then that, too, is relatively undefined, and perhaps infinite. The designer's difficulties are therefore twofold: understanding the problem and finding a solution.

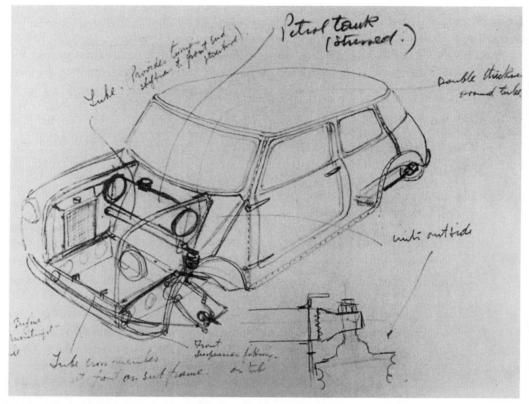

Figure 1.3   Generation: concept sketch for the Mini car by its designer Alec Issigonis

Often these two complementary aspects of design – problem and solution – have to be developed side-by-side. The designer makes a solution proposal and uses that to help understand what the problem 'really' is and what appropriate solutions might be like. The very first conceptualizations and representations of problem and solution are therefore critical to the kinds of searches and other procedures that will follow, and so to the final solution that will be designed.

The exploration of design solution-and-problem is also often done through early sketching of tentative ideas. It is necessary because normally there is no way of directly generating an 'optimum' solution from the information provided in the design brief. Quite apart from the fact that the client's brief to the designer may be rather vague, there will be a wide range of constraints and criteria to be satisfied, and probably no single objective that must be satisfied above all others, as suggested in the problem–solution 'exploration' in Figure 1.4.

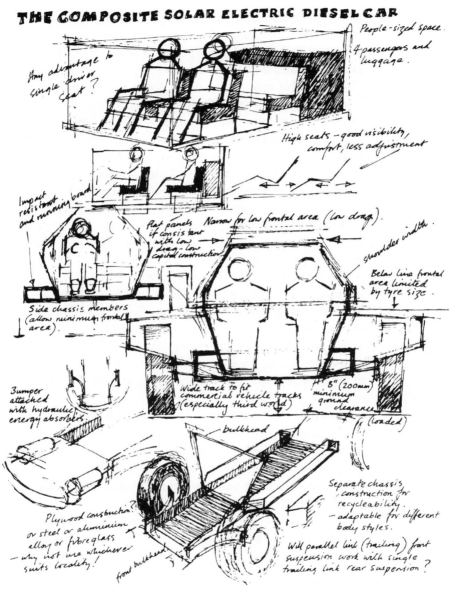

THE COMPOSITE SOLAR ELECTRIC DIESEL CAR

Figure 1.4 Exploration: an example of problem and solution being explored together for the Africar, a simple but robust automobile suitable for conditions in developing countries

Design Problems

Design problems normally originate as some form of problem statement provided to the designer by someone else, the client or the company management. These problem statements, normally called a design 'brief', can vary widely in their form and content. At one extreme, they might be something like the statement made by President Kennedy in 1961, setting a goal for the USA, 'before the end of the decade, to land a man on the moon and bring him back safely'. In this case, the goal was fixed, but the means of achieving it were very uncertain. The only constraint in the 'brief' was one of time: 'before the end of the decade'. The designers were given a completely novel problem, a fixed goal, only one constraint, and huge resources of money, materials and people. This is quite an unusual situation for designers to find themselves in!

At the other extreme is the example of the brief provided to the industrial designer Eric Taylor, for an improved pair of photographic darkroom forceps. According to Taylor, the brief originated in a casual conversation with the managing director of the photographic equipment company for which he worked, who said to him, 'I was using these forceps last night, Eric. They kept slipping into the tray. I think we could do better than that.' In this case, the brief implied a design modification to an existing product, the goal was rather vague – 'that [they] don't slip into the tray' – and the resources available to the designer would have been very limited for such a low-cost product. Taylor's redesign provided ridges on the handles of the forceps, to prevent them slipping against the side of the developing tray.

Somewhere between these extremes would fall the more normal kind of design brief. A typical example might be the following brief provided to the design department by the planning department of a company manufacturing plumbing fittings (source: Pahl and Beitz, 1999). It is for a domestic hot and cold water mixing tap that can be operated with one hand.

One-handed water mixing tap

Required: one-handed household water mixing tap with the following characteristics:

Throughput	10 l/min
Maximum pressure	6 bar
Normal pressure	2 bar
Hot water temperature	60 °C
Connector size	10 mm

Attention to be paid to appearance. The firm's trademark to be prominently displayed. Finished product to be marketed in two years' time. Manufacturing costs not to exceed DM30 each at a production rate of 3000 taps per month.

What all of these three examples of design problems have in common is that they set a *goal*, some *constraints* within which the goal must be achieved and some *criteria* by which a successful solution might be recognized. They do not specify what the solution will be, and there is no certain way of proceeding from the statement of the problem to a statement of the solution, except by 'designing'. Unlike some other kinds of problems, the person setting the problem does not know what the 'answer' is, but they will recognize it when they see it.

Even this last statement is not always true; sometimes clients do not 'recognize' the design solution when they see it. A famous example of early modern architecture was the 'Tugendhat House' in Brno, Czechoslovakia, designed in 1930 by Ludwig Mies van der Rohe. Apparently the client had approached the architect after seeing some of the rather more conventional houses that he had designed. According to Mies van der Rohe, when he showed the surprising, new design to the client, 'He wasn't very happy at first. But then we smoked some good cigars, ... and we drank some glasses of a good Rhein wine, ... and then he began to like it very much.'

So the solution that the designer generates may be something that the client 'never imagined might be possible', or perhaps even 'never realized was what they wanted'. Even a fairly precise problem statement gives no indication of what a solution *must* be. It is this uncertainty that makes designing such a challenging activity.

Ill-defined problems

The kinds of problems that designers tackle are regarded as 'ill-defined' or 'ill-structured', in contrast to well-defined or well-structured problems such as chess-playing, crossword puzzles or standard calculations. *Well-defined* problems have a clear goal, often one correct answer, and rules or known ways of proceeding that will generate an answer. The characteristics of *ill-defined* problems can be summarized as follows:

1. *There is no definitive formulation of the problem*. When the problem is initially set, the goals are usually vague, and many constraints and criteria are unknown. The problem context is often complex and messy, and poorly understood. In the course of problem-solving, temporary formulations of the problem may be fixed, but these are unstable and can change as more information becomes available.

2. *Any problem formulation may embody inconsistencies.* The problem is unlikely to be internally consistent; many conflicts and inconsistencies have to be resolved in the solution. Often, inconsistencies emerge only in the process of problem-solving.

3. *Formulations of the problem are solution-dependent.* Ways of formulating the problem are dependent upon ways of solving it; it is difficult to formulate a problem statement without implicitly or explicitly referring to a solution concept. The way the solution is conceived influences the way the problem is conceived.

4. *Proposing solutions is a means of understanding the problem.* Many assumptions about the problem and specific areas of uncertainty can be exposed only by proposing solution concepts. Many constraints and criteria emerge as a result of evaluating solution proposals.

5. *There is no definitive solution to the problem.* Different solutions can be equally valid responses to the initial problem. There is no objective true-or-false evaluation of a solution; but solutions are assessed as good or bad, appropriate or inappropriate.

Design problems are widely recognized as being ill-defined problems. It is usually possible to take some steps towards improving the initial definition of the problem, by questioning the client, collecting data, carrying out research, etc. There are also some rational procedures and techniques which can be applied in helping to solve ill-defined problems. But the designer's traditional approach, as suggested in some of the statements about ill-defined problems listed above, is to try to move fairly quickly to a potential solution, or set of potential solutions, and to use that as a means of further defining and understanding the problem.

Problem Structures

However, even when the designer has progressed well into the definition of a solution, difficulties in the problem structure may well still come to light. In particular, subsolutions can be found to be interconnected with each other in ways that form a 'pernicious', circular structure to the problem – e.g. a subsolution that resolves a particular subproblem may create irreconcilable conflicts with other subproblems.

An example of this pernicious problem structure was found in a study of housing design. The architects identified five decision areas, or subproblems, concerned with the directions of span of the roof and first floor joists, and the provision of load-bearing or nonload-bearing walls and partitions at ground- and first-floor levels. Making a decision in one area (say, direction of roof span) had implications for the first-floor partitions, and therefore the ground-floor partitions, which had implications for the direction of span of first-floor joists, and therefore for which of the external walls would have to be designed to be load-bearing. This not only had implications for the design of the external wall elevations, but also for the direction of span of the roof; and so one comes full circle back to the first decision area. This problem structure is shown diagrammatically in Figure 1.5, illustrating the circular structure that is often found in design problems.

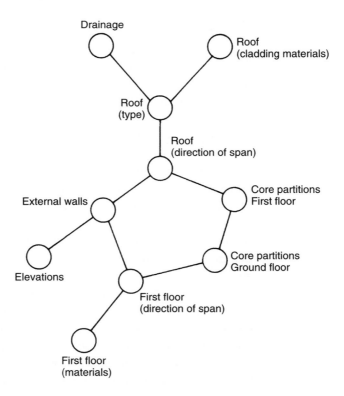

Figure 1.5
Problem structure found in a housing design problem

As part of the research study, the individual subsolution options in each decision area were separated out and the incompatible pairs of options identified. With this approach, it was possible to enumerate all the feasible solutions (i.e. sets of five options containing no incompatible pairs). There were found to be eight feasible solutions, and relative costings of each could indicate which would be the cheapest solution. This approach was later generalized into a new design method: AIDA, the Analysis of Interconnected Decision Areas (Luckman, 1984).

This example shows that a rigorous approach can sometimes be applied even when the problem appears to be ill-defined and the problem structure pernicious. This lends some support to those who argue that design problems are not always as ill-defined or ill-structured as they might appear to be. However, research into the behaviour of designers has shown that they will often treat a given problem *as though* it is ill-structured, even when it is presented as a well-structured problem, in order that they can create something innovative.

Research has also shown that designers often attempt to avoid cycling around the pernicious decision loops of design problems by making high-level strategic decisions about solution options. Having identified a number of options, the designer selects what appears to be the best one for investigation at a more detailed level; again, several options are usually evident, and again a choice is made. This results in what is known as a 'decision tree', with more and more branches opening from each decision point. An example is shown in Figure 1.6, based on a study of an engineer designing a 'carrying/fastening device' for attaching a backpack to a mountain bicycle.

This decision tree was derived from an experimental study in which the designer's progress was recorded over a two-hour period. The decision tree shows how higher-level strategic decisions (such as, in this case, positioning the device at either the front or rear wheel of the bicycle) gradually unfolded into lower-level implications and decisions, right down to details of screws, pins, etc. (Dwarakanath and Blessing, 1996).

The decision tree analysis of the design process perhaps implies that the result is the best possible design, if the best options are chosen at each level. However, a decision at any particular level may well turn out to be suboptimal in the light of subsequent options available at the other levels. For this reason, there is frequent backtracking up and down the levels of hierarchy in the design tree.

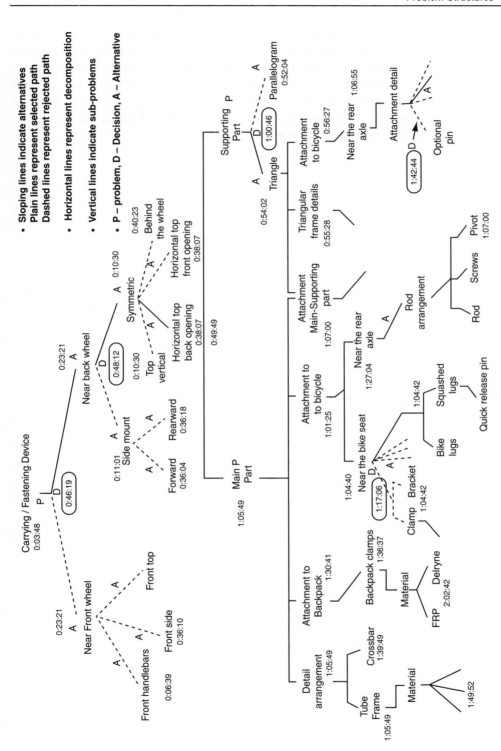

Figure 1.6 Decision tree derived from the design of a device for carrying a backpack on a bicycle

In Figure 1.6 this is confirmed by some of the 'time stamps' inserted at points within the tree, recording the time at which the designer considered the various alternatives and made decisions.

Resolving design problems by a 'top-down' approach is quite common, although sometimes a 'bottom-up' approach is used, starting with the lowest-level details and building up to a complete overall solution concept.

2 Design Ability

What Designers Say

The wish to design things is inherent in human beings, and 'design' is not something that has always been regarded as needing special abilities. It is only with the emergence and growth of industrial societies that the ability to design has become regarded as a specialized talent. But although there is so much design activity going on in the world, the ways in which people design are actually rather poorly understood. Design ability has been regarded as something that perhaps many people possess to some degree, but only a few people have a particular design 'talent'. However, there is now a growing body of knowledge about the nature of designing, about design ability and how to develop it and about the design process and how to improve it.

When designers are asked to discuss their abilities, and to explain how they work, a few common themes emerge. One theme is the importance of creativity and 'intuition' in design – even in engineering design. For example, the engineering designer Jack Howe has said:

> I believe in intuition. I think that's the difference between a designer and an engineer ... I make a distinction between engineers and engineering designers ... An engineering designer is just as creative as any other sort of designer.

Engineering Design Methods: Strategies for Product Design   N. Cross
© 2008 John Wiley & Sons, Ltd

Some rather similar comments have been made by the industrial designer Richard Stevens:

A lot of engineering design is intuitive, based on subjective thinking. But an engineer is unhappy doing this. An engineer wants to test; test and measure. He's been brought up this way and he's unhappy if he can't prove something. Whereas an industrial designer ... is entirely happy making judgements which are intuitive.

Another theme that emerges from what designers say about their abilities is based on the recognition that problems and solutions in design are closely interwoven; that 'the solution' is not always a straightforward answer to 'the problem'. For example, the furniture designer Geoffrey Harcourt commented on one of his creative designs like this:

As a matter of fact, the solution that I came up with wasn't a solution to the problem at all. I never saw it as that ... But when the chair was actually put together (it) in a way quite well solved the problem, but from a completely different angle, a completely different point of view.

A third common theme to emerge is the need to use sketches, drawings or models of various kinds as a way to explore the problem and solution together. The conceptual thinking processes of the designer seem to be based on the development of ideas through their external expression in sketches. As the engineer–architect Santiago Calatrava said:

To start with you see the thing in your mind and it doesn't exist on paper and then you start making simple sketches and organizing things and then you start doing layer after layer ... it is very much a dialogue.

This 'dialogue' occurs through the designer's perception of the sketched concepts, and reflection on the ideas that they represent and their implications for the resolution of the problem. The designer responds to the perceptions, reflections and implications, and so the 'dialogue' between internal mental processes and external representations continues.

The quotations above are taken from interviews conducted with a number of successful and eminent designers by Davies (1985) and Lawson (1994). The designers' comments support some of the hypotheses that have emerged from more objective observational studies of designers at work, and other research that has been conducted into the nature of design. Some of this research reflects the view that designers have a particular, 'designerly' way of thinking and working (Cross, 2007).

How Designers Think

In an experimental research study, Lawson (1984) compared the ways in which designers (in this case architects) and scientists solved the same problem. The scientists tended to use a strategy of systematically trying to understand the problem, in order to look for underlying rules which would enable them to generate an optimum solution. In contrast, the designers tended to make initial explorations and then suggest a variety of possible solutions until they found one that was good, or at least satisfactory. The evidence from the experiments suggested that scientists problem-solve by analysis, whereas designers problem-solve by synthesis; scientists use 'problem-focused' strategies and designers use 'solution-focused' strategies.

Some other studies have also suggested that designers tend to use conjectures about solution concepts as the means of developing their understanding of the problem. Darke (1984) found that designers impose a 'primary generator' onto the problem, in order to narrow the search space and generate early solution concepts. This 'primary generator' is usually based on a tightly restricted set of constraints or solution possibilities derived from the design problem. Since the 'problem' cannot be fully understood in isolation from consideration of the 'solution', it is natural that solution conjectures should be used as a means of helping to explore and understand the problem formulation. Making sketches of solution concepts is one way that helps the designer to identify their consequences, and to keep the problem exploration going, in what Schön (1983) called the 'reflective conversation with the situation' that is characteristic of design thinking.

Drawing and sketching have been used in design for a long time, certainly since long before the Renaissance, but the period since that time has seen a massive growth in the use of drawings, as designed objects have become more complex and more novel. Many of Leonardo da Vinci's drawings of machines and inventions from the Renaissance period show one of the key aspects of design drawings, in terms of their purpose of communicating to someone else how a new product should be built, and also how it should work. Some of Leonardo's design drawings also show how a drawing can be not only a communication aid, but also a thinking and reasoning aid.

For example, Leonardo's sketches for the design of fortifications (Figure 2.1) show how he used sight-lines and missile trajectories as lines to set up the design of the fortifications, and how his design thinking was assisted by drawing. In such drawings we see how the sketch can help the designer to consider many aspects at once – we see plans, elevations, details, trajectory lines, all being drawn together and thus all being thought about, reasoned about, all together.

Half a millennium later, we still see designers using essentially similar types of sketches to aid their design thinking. The early concept sketches for a house design by the contemporary architect Charles Moore (Figure 2.2) show similar kinds of representations as those used by Leonardo – plan, elevation and section all being considered together with considerations of structure and calculations of dimensions and areas.

What might we learn about the nature of design thinking from looking at examples of what designers sketch? One thing that seems to appear is that sketches enable designers to handle different levels of abstraction simultaneously. Clearly this is something important in the design process. We see that designers think about the overall concept and at the same time think about detailed aspects of the implementation of that concept. Obviously not *all* of the detailed aspects are considered early on, because if they could do that, designers could go straight to the final set of detailed drawings. So they use the concept sketch to identify and then to reflect upon *critical* details, particular details that they realize might hinder, or somehow significantly influence the final implementation of the complete design. This implies that, although there is a hierarchical structure of decisions, from overall concept to details, designing is not a strictly hierarchical process; in the early stages of design, the designer moves freely between different levels of detail.

The identification of critical details is part of a more general facility that sketches provide, which is that they enable identification and recall of relevant knowledge. As the architect Richard MacCormac has said about designing, 'What you need to know about the problem only becomes apparent as you're trying to solve it.' There is a massive amount of information that *may* be relevant,

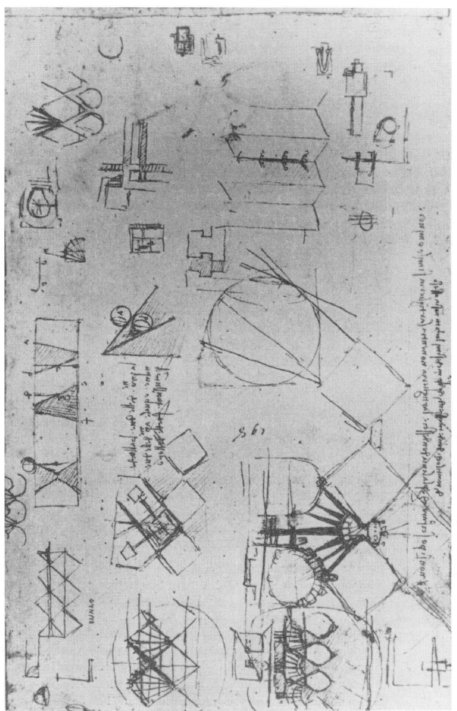

Figure 2.1 Sketch design studies for town fortifications by Leonardo da Vinci (c. 1500)

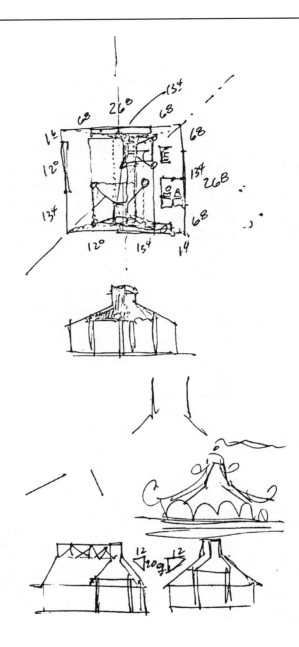

Figure 2.2
Sketch design
drawings for a small
house by the
architect Charles
Moore (*c.* 1960)

not only to *all* the possible solutions for a design problem, but simply
to *any* possible solution. And any possible solution in itself creates
the unique circumstances in which these large bodies of information
interact, probably in unique ways for any one possible solution.
So these large amounts of information and knowledge need to be
brought into play in a selective way, being selected only when they

become relevant, as the designer considers the implications of the solution concept as it develops.

Because the design problem is itself ill-defined and ill-structured, a key feature of design sketches is that they assist problem structuring through the making of solution attempts. Sketches incorporate not only drawings of tentative solution concepts but also numbers, symbols and text, as designers relate what they know of the design problem to what is emerging as a solution. Sketching enables exploration of the problem space and the solution space to proceed together, assisting the designer to converge on a matching problem–solution pair. Problem and solution co-evolve in the design process.

Designers' use of sketches therefore gives us some considerable insight into the nature of design thinking and the resolution of design problems. These problems cannot be stated sufficiently explicitly such that solutions can be derived directly from them. The designer has to take the initiative in finding a problem starting point and suggesting tentative solution areas. Problem and solution are then both developed in parallel, sometimes leading to a creative redefinition of the problem, or to a solution that lies outside the boundaries of what was previously assumed to be possible.

Solution-focused strategies are therefore perhaps the best way of tackling design problems, which are by nature ill-defined. In order to cope with the uncertainty of ill-defined problems, the designer has to have the self-confidence to define, redefine and change the problem as given, in the light of solutions that emerge in the very process of designing. People who prefer the certainty of structured, well-defined problems will never appreciate the delight of being a designer!

Learning to Design

An appropriate use of the 'solution-focused' approach to design is something that seems to develop with experience. Experienced designers are able to draw on their knowledge of previous exemplars in their field of design, and they also seem to have learned the value of rapid problem exploration through solution conjecture. In comparison, novice designers can often become bogged down in attempts to understand the problem before they start generating solutions. For them, gathering data about the problem is sometimes just a substitute activity for actually doing any design work.

However, novice designers are also frequently found to become 'fixated' on particular solution concepts. Early solution concepts are often found to be less than satisfactory, as problem exploration continues. But novice designers (and sometimes more experienced ones) can be loath to discard the concept and return to a search for a better alternative. Instead, they try laboriously to 'design-out' the imperfections in the concept, producing slight improvements until something workable, but perhaps far from ideal, is attained. Sometimes it can be much more productive to start afresh with a new design concept.

Another difference between novices and experts is that novices will often pursue a 'depth-first' approach to a problem – sequentially identifying and exploring subsolutions in depth, and amassing a number of partial subsolutions that then somehow have to be amalgamated and reconciled, in a 'bottom-up' process. Experts usually pursue predominantly 'breadth-first' and top-down strategies, as recorded in the example of the expert designer's decision tree in Figure 1.6 (Chapter 1).

Experienced designers, like any skilled professionals, can make designing seem easy and 'intuitive'. Because skilled design in practice therefore often appears to proceed in a rather ad-hoc and unsystematic way, some people claim that learning a systematic process does not actually help student designers. However, studies in design education have shown that a systematic approach can be helpful to students. For example, Radcliffe and Lee (1989) found that the use of more 'efficient' design processes (following closer to an 'ideal' sequence) correlated positively with both the quantity and the quality of students' design results. Other studies have tended to confirm this.

From studies of a number of engineering designers, of varying degrees of experience and with varying exposures to education in systematic design processes, Fricke (1996) found that designers following a 'flexible-methodical procedure' tended to produce good solutions. These designers worked reasonably efficiently and followed a fairly logical procedure, whether or not they had been educated in a systematic approach. In comparison, designers either with a too-rigid adherence to a systematic procedure (behaving 'unreasonably methodical'), or with very unsystematic approaches, produced mediocre or poor design solutions. Successful designers (ones producing better quality solutions) tended to be those who:

- clarified requirements, by asking sets of related questions which focused on the problem structure;

- actively searched for information, and critically checked given requirements;

- summarized information on the problem formulation into requirements and partially prioritized them;

- did not suppress first solution ideas; they held on to them, but returned to clarifying the problem rather than pursuing initial solution concepts in depth;

- detached themselves during conceptual design stages from fixation on early solution concepts;

- produced variants but limited the production and kept an overview by periodically assessing and evaluating in order to reduce the number of possible variants.

The key to successful design therefore seems to be the effective management of the dual exploration of both the 'problem space' and the 'solution space'.

Designing is a form of skilled behaviour. Learning any skill usually relies on controlled practice and the development of techniques. The performance of a skilled practitioner appears to flow seamlessly, adapting the performance to the circumstances without faltering. But learning is not the same as performing, and underneath skilled performance lies mastery of technique and procedure.

3 *The Design Process*

Descriptive Models

There have been many attempts to draw up maps or models of the design process. Some of these models simply *describe* the sequences of activities that typically occur in designing; other models attempt to *prescribe* a better or more appropriate pattern of activities.

Descriptive models of the design process usually identify the significance of generating a solution concept early in the process, thus reflecting the 'solution-focused' nature of design thinking. This initial solution 'conjecture' is then subjected to analysis, evaluation, refinement and development. Sometimes, of course, the analysis and evaluation show up fundamental flaws in the initial conjecture and it has to be abandoned, a new concept generated and the cycle started again. The process is *heuristic*: using previous experience, general guidelines and 'rules of thumb' that lead in what the designer hopes to be the right direction, but with no absolute guarantee of success.

In Chapter 1, I developed a simple descriptive model of the design process, based on the essential activities that the designer performs. The endpoint of the process is the *communication* of a design, ready for manufacture. Prior to this, the design proposal is subject to *evaluation* against the goals, constraints and criteria of the design brief. The proposal itself arises from the *generation* of a concept by the designer, usually after some initial *exploration* of the ill-defined problem space. Putting these four activity types in their natural sequence, we have a simple four-stage model of the design process consisting of: exploration; generation; evaluation; and communication.

Engineering Design Methods: Strategies for Product Design   N. Cross
© 2008 John Wiley & Sons, Ltd

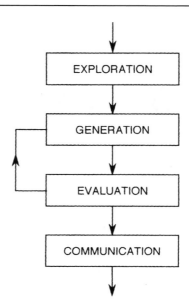

Figure 3.1
Simple four-stage
model of the design
process

This simple four-stage model is shown diagrammatically in Figure 3.1. Assuming that the evaluation stage does not always lead directly onto the communication of a final design, but that sometimes a new, more satisfactory concept has to be chosen, an iterative feedback loop is shown from the evaluation stage to the generation stage.

Models of the design process are often drawn in this flow-diagram form, with the development of the design proceeding from one stage to the next, but with feedback loops showing the iterative returns to earlier stages which are frequently necessary. For example, French (1999) developed a more detailed model of the design process, shown in Figure 3.2, based on the following activities:

- analysis of problem

- conceptual design

- embodiment of schemes

- detailing.

In the diagram, the circles represent stages reached, or outputs, and the rectangles represent activities, or work in progress.

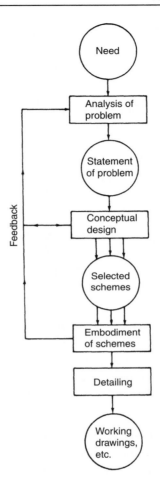

Figure 3.2
French's model of
the design process

The process begins with an initial statement of a 'need', and the first design activity is 'analysis of the problem'. French suggests that:

The analysis of the problem is a small but important part of the overall process. The output is a statement of the problem, and this can have three elements,

1. *A statement of the design problem proper*

2. *Limitations placed upon the solution, e.g. codes of practice, statutory requirements, customers' standards, date of completion, etc.*

3. *The criterion of excellence to be worked to.*

These three elements correspond to the goals, constraints and criteria of the design brief.

The activities that follow, according to French, are then:

Conceptual design

This phase ... takes the statement of the problem and generates broad solutions to it in the form of schemes. It is the phase that makes the greatest demands on the designer, and where there is the most scope for striking improvements. It is the phase where engineering science, practical knowledge, production methods, and commercial aspects need to be brought together, and where the most important decisions are taken.

Embodiment of schemes

In this phase the schemes are worked up in greater detail and, if there is more than one, a final choice between them is made. The end product is usually a set of general arrangement drawings. There is (or should be) a great deal of feedback from this phase to the conceptual design phase.

Detailing

This is the last phase, in which a very large number of small but essential points remains to be decided. The quality of this work must be good, otherwise delay and expense or even failure will be incurred; computers are already reducing the drudgery of this skilled and patient work and reducing the chance of errors, and will do so increasingly.

These activities are typical of conventional engineering design. Figures 3.3–3.5 illustrate the type of work that goes on in each stage. The illustrations are examples from the design of a small concrete mixer (Hawkes and Abinett, 1984). *Conceptual design* is

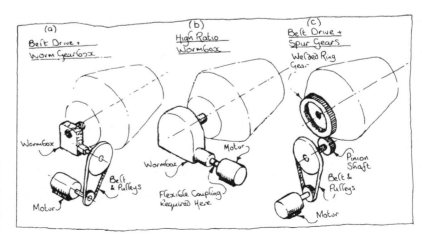

Figure 3.3
Conceptual design: alternatives for the drive connection in a small concrete mixer

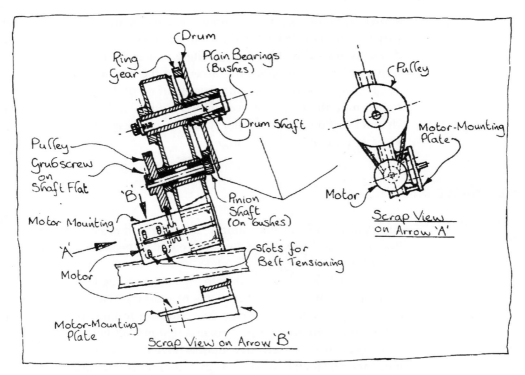

Figure 3.4   Embodiment design: one concept developed in more detail

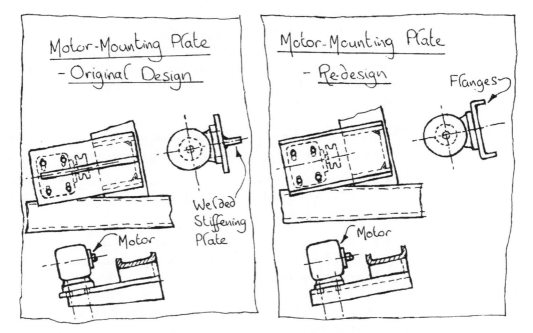

Figure 3.5   Detail design: redesign of a mounting plate to resist vibration

shown in Figure 3.3, where three alternatives are proposed for the drive connection from the motor to the mixing drum. *Embodiment design* is shown in Figure 3.4, where concept (c) is developed in terms of how to support and assemble the motor, drum, pulleys, etc. Figure 3.5 shows a small example of *detail design*, in which the motor mounting plate is redesigned from a welded T-shape to a channel section U-shape, after tests of a prototype found excessive vibration occurring in the original.

Prescriptive Models

As well as models that simply describe a more-or-less conventional, heuristic process of design, there have been several attempts at building prescriptive models of the design process. These latter models are concerned with trying to persuade or encourage designers to adopt improved ways of working. They usually offer a more *algorithmic*, systematic procedure to follow, and are often regarded as providing a particular *design methodology*.

Many of these prescriptive models have emphasized the need for more analytical work to precede the generation of solution concepts. The intention is to try to ensure that the design problem is fully understood, that no important elements of it are overlooked and that the 'real' problem is identified. There are plenty of examples of excellent solutions to the wrong problem!

These models have therefore tended to suggest a basic structure to the design process of analysis–synthesis–evaluation. These stages were defined by Jones (1984) in an early example of a systematic design methodology as follows:

- *Analysis. Listing of all design requirements and the reduction of these to a complete set of logically related performance specifications.*

- *Synthesis. Finding possible solutions for each individual performance specification and building up complete designs from these with least possible compromise.*

- *Evaluation. Evaluating the accuracy with which alternative designs fulfil performance requirements for operation, manufacture and sales before the final design is selected.*

This may sound very similar to a conventional design process, but the emphases here are on performance specifications logically

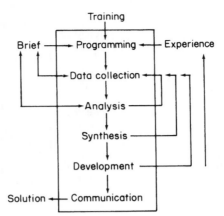

Figure 3.6
Archer's model of
the design process

derived from the design problem, generating several alternative design concepts by building up the best subsolutions and making a rational choice of the best of the alternative designs. Such apparently sensible, rational procedures are not always followed in conventional design practice.

A more detailed prescriptive model was developed by Archer (1984), and is summarized in Figure 3.6. This includes interactions with the world outside of the design process itself, such as inputs from the client, the designer's training and experience, other sources of information, etc. The output is, of course, the communication of a specific solution. These various inputs and outputs are shown as external to the design process in the flow diagram, which also features many feedback loops.

Within the design process, Archer identified six types of activity:

- *Programming – establish crucial issues; propose a course of action.*

- *Data collection – collect, classify and store data.*

- *Analysis – identify subproblems; prepare performance (or design) specifications; reappraise proposed programme and estimate.*

- *Synthesis – prepare outline design proposals.*

- *Development – develop prototype design(s); prepare and execute validation studies.*

- *Communication – prepare manufacturing documentation.*

Archer summarized this process as dividing into three broad phases: analytical, creative and executive (Figure 3.7). He suggested that:

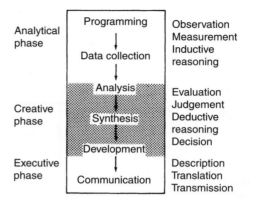

Figure 3.7
Archer's three-phase
summary model of
the design process

One of the special features of the process of designing is that the analytical phase with which it begins requires objective observation and inductive reasoning, while the creative phase at the heart of it requires involvement, subjective judgement, and deductive reasoning. Once the crucial decisions are made, the design process continues with the execution of working drawings, schedules, etc., again in an objective and descriptive mood. The design process is thus a creative sandwich. The bread of objective and systematic analysis may be think or thin, but the creative act is always there in the middle.

Some much more complex models have been proposed, but they often tend to obscure the general structure of the design process by swamping it in the fine detail of the numerous tasks and activities that are necessary in all practical design work. A reasonably comprehensive model that still retains some clarity is that offered by Pahl and Beitz (1999) (Figure 3.8). It is based on the following design stages:

- *Clarification of the task: Collect information about the requirements to be embodied in the solution and also about the constraints.*

- *Conceptual design: Establish function structures; search for suitable solution principles; combine into concept variants.*

- *Embodiment design: Starting from the concept, the designer determines the layout and forms and develops a technical product or system in accordance with technical and economic considerations.*

- *Detail design: Arrangement, form, dimensions and surface properties of all the individual parts finally laid down; materials specified; technical and economic feasibility re-checked; all drawings and other production documents produced.*

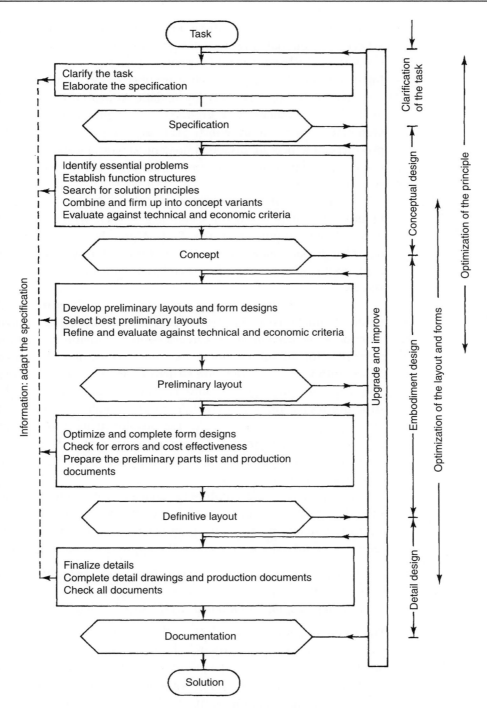

Figure 3.8 Pahl and Beitz's model of the design process

Considerable work on these kinds of models and on other aspects of rationalizing the design process has been done in Germany. The professional engineers' society, Verein Deutscher Ingenieure (VDI), has produced a number of 'VDI-Guidelines' in this area, including VDI 2221 'Systematic Approach to the Design of Technical Systems and Products'. This guideline suggests a systematic approach in which 'The design process, as part of product creation, is subdivided into general working stages, making the design approach transparent, rational and independent of a specific branch of industry.'

The structure of this general approach to design is shown in Figure 3.9, and is based on seven stages, each with a particular output. The output from the first stage, the specification, is regarded as particularly important, and is constantly reviewed and kept up to date, and used as a reference in all the subsequent stages.

The second stage of the process consists of determining the required functions of the design, and producing a diagrammatic function structure. In stage three a search is made for solution principles for all subfunctions, and these are combined in accordance with the overall function structure into a principal solution. This is divided, in stage four, into realizable modules and a module structure representing the breakdown of the solution into fundamental assemblies. Key modules are developed in stage five into a set of preliminary layouts. These are refined and developed in stage six into a definitive layout, and the final product documents are produced in stage seven.

In the guideline it is emphasized that several solution variants should be analysed and evaluated at each stage, and that there is a lot more detail in each stage than is shown in the diagram. The following words of warning about the approach are also given:

> *It is important to note that the stages do not necessarily follow rigidly one after the other. They are often carried out iteratively, returning to preceding ones, thus achieving a step-by-step optimization.*

The VDI-Guideline follows a general systematic procedure of first analysing and understanding the problem as fully as possible, then breaking this into subproblems, finding suitable subsolutions and combining these into an overall solution. The procedure is shown diagrammatically in Figure 3.10.

This kind of procedure has been criticized in the design world because it seems to be based on a problem-focused rather than a

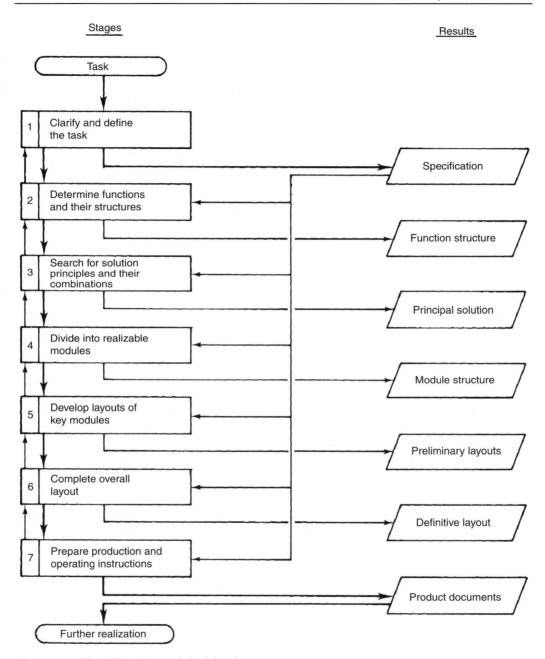

Figure 3.9 The VDI 2221 model of the design process

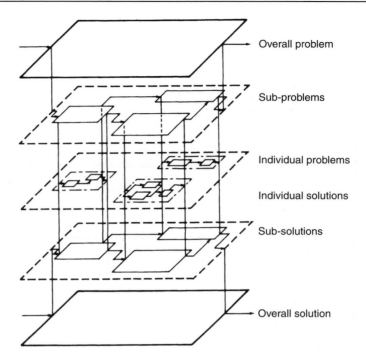

Overall problem

Sub-problems

Individual problems

Individual solutions

Sub-solutions

Overall solution

Figure 3.10
The VDI 2221 model
of development
from problem to
solution

solution-focused approach. It therefore runs counter to designers'
traditional ways of thinking.

A more radical model of the design process, which recognizes
the solution-focused nature of design thinking, has been suggested
by March (1984) (Figure 3.11). He argued that the two conven-
tionally understood forms of reasoning, inductive and deductive,
only apply logically to the evaluative and analytical types of activity
in design. But the type of activity that is most particularly associated
with design is that of synthesis, for which there is no commonly
acknowledged form of reasoning. March drew on the work of the
philosopher Peirce to identify this missing concept of 'abductive'
reasoning. According to Peirce,

> Deduction proves that something *must be*; induction shows
> that something *actually is* operative; abduction suggests that
> something *may be*.

It is this hypothesizing of what *may* be, the act of synthesis, that is
central to design. Because it is the kind of thinking by which designs
are generated or produced, March prefers to call it 'productive'
reasoning. Thus his model for a rational design process is a 'PDI
model': production – deduction – induction.

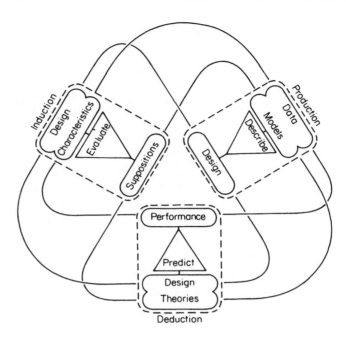

Figure 3.11
March's model of
the design process

In this model, the first phase, of *productive* reasoning, draws on a preliminary statement of requirements, and some presuppositions about solution types in order to produce, or describe, a design proposal. From this proposal and established theory (e.g. engineering science) it is possible *deductively* to analyse, or predict, the performance of the design. From these predicted performance characteristics it is possible *inductively* to evaluate further suppositions or possibilities, leading to changes or refinements in the design proposal.

An Integrative Model

Certainly it seems that, in most design situations, it is not possible, or relevant, to attempt to analyse 'the problem' *ab initio* and in abstract isolation from solution concepts; the designer explores and develops problem and solution together. Although there may be some logical progression from problem to subproblems and from subsolutions to solution, there is a symmetrical, commutative relationship between problem and solution, and between subproblems and subsolutions, as illustrated in Figure 3.12. This model attempts to capture the

Figure 3.12
Symmetrical
relationships of
problem/sub-
problems/sub-
solutions/solution
in design

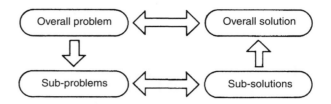

Figure 3.12 Symmetrical relationships of problem/sub-problems/sub-solutions/solution in design

essential nature of the design process, in which understanding of the problem and of the solution develops together, or co-evolves. There is a constant transfer of the designer's attention backwards and forwards between the 'problem space' (left-hand side of the model) and the 'solution space' (right-hand side of the model). The model also attempts to recognize that there is an expected pattern of progression in the design process, from a given problem to a proposed solution. There is therefore assumed to be a general, anticlockwise direction of movement in the model, from top left around to top right, but with substantial periods of iterative activity, going to-and-fro between problem and solution, subproblems and subsolutions.

Part Two
Doing Design

4 New Design Procedures

Systematic Procedures

There may be differences in their preferred models, but the proponents of new models of the design process all agree that there is a need to improve on traditional ways of working in design.

There are several reasons for this concern to develop new design procedures. One is the increasing complexity of modern design. A great variety of new demands is increasingly being made on designers, such as the new materials and devices (e.g. electronics) that become available and the new problems that are presented to designers. Many of the products and machines to be designed today have never existed before, and so the designer's previous experience may well be irrelevant and inadequate for these tasks. Therefore a new, more systematic approach is needed, it is argued.

A related part of the complexity of modern design is the need to develop team work, with many specialists collaborating in and contributing to the design. To help coordinate the team, it is necessary to have a clear, organized approach to design, so that specialists' contributions are made at the right point in the process. Dividing the overall problem into subproblems in a systematic procedure also means that the design work itself can be subdivided and allocated to appropriate team members.

As well as being more complex, modern design work often has very high risks and costs associated with it. For example, many products are designed for mass manufacture, and the costs of setting up the manufacturing plant, buying in raw materials, and so on, are so high that the designer cannot afford to make mistakes: the design must be absolutely right before it goes into production. This means

Engineering Design Methods: Strategies for Product Design N. Cross
© 2008 John Wiley & Sons, Ltd

that any new product must have been through a careful process of design. Other kinds of large, one-off designs, such as chemical process plants, or complex products such as aeroplanes, also have to have a very rigorous design process to try to ensure their safe operation and avoid the catastrophic consequences of failure.

Finally, there is a more general concern with trying to improve the efficiency of the design process. In some industries there is a pressing need to ensure that the lead-time necessary to design a new product is kept to a minimum. In all cases, it is desirable to try to avoid the mistakes and delays that often occur in conventional design procedures. The introduction of computers already offers one way of improving the efficiency of the design process, and is also in itself an influence towards more systematic ways of working.

Design Methods

One of the most significant aspects of this concern to improve the design process has been the development of new design methods. In a sense, any identifiable way of working, within the context of designing, can be considered to be a design method. The most common design method can be called the method of 'design-by-drawing'. That is to say, most designers rely extensively on drawing as their main aid to designing.

Design methods can, therefore, be any procedures, techniques, aids or 'tools' for designing. They represent a number of distinct kinds of activities that the designer might use and combine into an overall design process. Although some design methods can be the conventional, normal procedures of design, such as drawing, there has been a substantial growth in new, unconventional procedures that are more usually grouped together under the name of 'design methods'.

The main intention of these new methods is that they attempt to bring rational procedures into the design process. It sometimes seems that some of these new methods can become *over*-formalized, or can be merely fancy names for old, commonsense techniques. They can also appear to be *too* systematic to be useful in the rather messy and often hurried world of the design office. For these kinds of reasons, many designers are still mistrustful of the whole idea of 'design methods'.

The counterarguments to that view are based on the reasons for adopting systematic procedures, outlined above. For instance, many modern design projects are too complex to be resolved satisfactorily by the old, conventional methods. There are also too many errors made with conventional ways of working, and they are not very useful where team work is necessary. Design methods try to overcome these kinds of problems, and, above all, they try to ensure that a better product results from the new design process. They can also be good practice methods for student designers, offering a training in certain ways of thinking and proceeding in design.

Some design methods are new inventions of rational procedures, some are adapted from operational research, decision theory, management sciences or other sources, and some are simply extensions or formalizations of the informal techniques that designers have always used. For example, the informal methods of looking up manufacturers' catalogues or seeking advice from colleagues might be formalized into an 'information search' method; or informal procedures for saving costs by detailed redesigning of a component can be formalized into a 'value analysis' method. Different design methods have different purposes and are relevant to different aspects of and stages in the design process.

The new methods tend to have two principal features in common. One is that they *formalize* certain procedures of design, and the other is that they *externalize* design thinking. Formalization is a common feature of design methods because they attempt to avoid the occurrence of oversights, of overlooked factors in the design problem, of the kinds of errors that occur with informal methods. The process of formalizing a procedure also tends to widen the approach that is taken to a design problem and to widen the search for appropriate solutions; it encourages and enables you to think beyond the first solution that comes into your head.

This is also related to the other general aspect of design methods, that they externalize design thinking, i.e. they try to get your thoughts and thinking processes out of your head and into the charts and diagrams that commonly feature in design methods. This externalizing is a significant aid when dealing with complex problems, but it is also a necessary part of team work, i.e. providing means by which all the members of the team can see what is going on and can contribute to the design process. Getting a lot of systematic work out of your head and onto paper also means that your mind can be more free to pursue the kind of thinking at which it is best: intuitive and imaginative thinking.

Design methods therefore are *not* the enemy of creativity, imagination and intuition. Quite the contrary: they are perhaps more likely to lead to novel design solutions than the informal, internal and often incoherent thinking procedures of the conventional design process. Some design methods are, indeed, techniques specifically for aiding creative thought. In fact, the general body of design methods can be classified into two broad groups: creative methods and rational methods.

Creative Methods

There are several design methods which are intended to help stimulate creative thinking. In general, they work by trying to increase the flow of ideas, by removing the mental blocks that inhibit creativity or by widening the area in which a search for solutions is made.

Brainstorming   The most widely known creative method is brainstorming. This is a method for generating a large number of ideas, most of which will subsequently be discarded, but with perhaps a few novel ideas being identified as worth following up. It is normally conducted as a small group session of about four to eight people.

The group of people selected for a brainstorming session should be diverse. It should not just be experts or those knowledgeable in the problem area, but should include a wide range of expertise and even laypeople if they have some familiarity with the problem area. The group must be nonhierarchical, although one person does need to take an organizational lead.

The role of the group leader in a brainstorming session is to ensure that the format of the method is followed, and that it does not just degenerate into a round-table discussion. An important prior task for the leader is to formulate the problem statement used as a starting point. If the problem is stated too narrowly, then the range of ideas from the session may be rather limited. On the other hand, a very vague problem statement leads to equally vague ideas, which may be of no practical use. The problem can often be usefully formulated as a question, such as 'How can we improve on X?'

In response to the initial problem statement, the group members are asked to spend a few minutes, in silence, writing down the first ideas that come into their heads. It is a good idea if each member has a pile of small record cards on which to write these and subsequent ideas. The ideas should be expressed succinctly, and written one per card.

The next, and major, part of the session is for each member of the group, in turn, to read out one idea from his or her set. The most important rule here is that *no criticism is allowed* from any other member of the group. The usual responses to unconventional ideas, such as 'That's silly' or 'That will never work', kill off spontaneity and creativity. At this stage, the feasibility or otherwise of any idea is not important; evaluation and selection will come later.

What each group member should do in response to every other person's idea is to try to build on it, to take it a stage further, to use it as a stimulus for other ideas, or to combine it with his or her own ideas. For this reason, there should be a short pause after each idea is read out, to allow a moment for reflection and for writing down further new ideas. However, the session must not become too stilted; the atmosphere should be relaxed and free-wheeling. A brainstorming session should also be fun: humour is often an essential ingredient of creativity.

The group session should not last more than about 20–30 minutes, or should be wound up when no more new ideas are forthcoming. The group leader, or someone else, then collects all the cards and spends a separate period evaluating the ideas. A useful aid to this evaluation is to sort or classify the ideas into related groups; this in itself often suggests further ideas, or indicates the major types of idea that there appear to be. If principal solution areas and one or two novel ideas result from a brainstorming session then it will have been worthwhile.

Participating in a brainstorming session is rather like playing a party game; and like a party game it only works well when everyone sticks to the rules. In fact, all design methods only work best when they are followed with some rigour, and not in a sloppy or half-hearted fashion. The essential rules of brainstorming are:

- No criticism is allowed during the session.

- A large quantity of ideas is wanted.

- Seemingly crazy ideas are quite welcome.

- Keep all ideas short and snappy.

- Try to combine and improve on the ideas of others.

Example: container lock   This example shows how brainstorming can be applied to the task of creating a new solution to an old problem: the locking of containers (the large goods containers transported by lorries). The conventional solution is a padlock, but then the key for the padlock also has to be either transported together with the container (hence presenting an obvious security problem) or sent separately to the recipient (possibly getting lost). In practice, it seems that most container padlocks are 'opened' with a bolt-cutter, because no-one can find the key!

A short brainstorming session was held to generate ideas for solving this problem. The problem was stated as: 'Provide a means of securing containers that is tamper-proof but easy to open.' Within a few minutes, the following ideas were generated:

- incorporate an electronic code – fax the code to the recipient;

- combination lock;

- time lock;

- clasps welded together;

- a locked bolt that is easily cut to open it;

- padlocks with master keys retained by regular customers;

- giant stapler and staple-remover;

- ceramic bolt that can be smashed;

- glass bolt that sounds alarm when smashed;

- lorry driver swallows the key;

- a 'puzzle' lock that can only be opened by a very skilled person.

Some of these are fairly 'obvious' ideas, but getting them out of your head can sometimes seem to free the mental space for other ideas to come. Others are 'crazy' ideas, such as the lorry driver swallowing the key; in such a case, everyone knows where the key is, but has to wait a couple of days before it can be recovered! (Another sort of 'time lock', as the proposer explained!) There is also an example in the list of one idea building upon another: the 'glass bolt that sounds an alarm when smashed' was a response to the 'ceramic bolt' idea, but based also on fire alarm buttons that are activated by smashing the glass cover.

In reviewing this list of ideas several novel concepts come to mind, but perhaps most appealing is the simplicity of adapting what is already the unofficial but conventional solution: to cut the bolt off. A bolt could be made such that it was designed to be cut off. Made in two sections, the parts of the bolt would be pushed together to secure the container, and could only be opened by being cut. Colour coding and numbering each bolt would mean that it could not be replaced in transit, and if it was cut open then this would be obvious. Such simple but secure bolts would be cheaper than conventional padlocks. The 'Oneseal' disposable container lock was designed on these principles.

Synectics

Creative thinking often draws on analogical thinking, on the ability to see parallels or connections between apparently dissimilar topics. The role of humour is again relevant, since most jokes depend for their effect on the unexpected transfer or juxtaposition of concepts from one context to another, or what Koestler called the 'bisociation' of ideas. Bisociation plays a fundamental role in creativity.

The use of analogical thinking has been formalized in a creative design method known as 'synectics'. Like brainstorming, synectics is a group activity in which criticism is ruled out, and the group members attempt to build, combine and develop ideas towards a creative solution to the set problem. Synectics is different from brainstorming in that the group tries to work collectively towards a particular solution, rather than generating a large number of ideas. A synectics session is much longer than brainstorming, and much more demanding. In a synectics session, the group is encouraged to use particular types of analogy, as follows:

Direct analogies

These are usually found by seeking a biological solution to a similar problem. For example, Brunel's observation of a shipworm forming a tube for itself as it bored through timber is said to have led him to the idea of a caisson for underwater constructions; 'Velcro' fastening was designed on an analogy with plant burrs.

Personal analogies

The team members imagine what it would be like to use oneself as the system or component that is being designed. For example, what would it feel like to be a motorcar suspension unit; how would I operate if I were a computerized filing system?

Symbolic analogies Here, poetic metaphors and similes are used to relate aspects of one thing with aspects of another. For example, the 'friendliness' of a computer, the 'head' and 'claw' of a hammer, a 'tree' of objectives, the 'Greek key pattern' of a housing layout.

Fantasy analogies These are 'impossible' wishes for things to be achieved in some 'magical' way. For example, 'What we really want is a doorkeeper who recognizes each system user'; 'We need the bumps in the road to disappear beneath the wheels'.

A synectics session starts with the 'problem as given': the problem statement as presented by the client or company management. Analogies are then sought that help to 'make the strange familiar', i.e. expressing the problem in terms of some more familiar (but perhaps rather distant) analogy. This leads to a conceptualization of the 'problem as understood': the key factor or elements of the problem that need to be resolved, or perhaps a complete reformulation of the problem. The problem as understood is then used to guide the use of analogies again, but this time to 'make the familiar strange'. Unusual, creative analogies are sought, which may lead to novel solution concepts. The analogies are used to open up lines of development which are pursued as hard and as imaginatively as possible by the group.

Example: forklift truck A design team looking for new versions of a company's forklift trucks focused on the problem area of using such trucks in warehouses for the stacking and removal of palletted goods. Conventional forklift trucks have to face head-on to the stacks in order to place and lift the pallets, and then be manoeuvred again within the aisle between the stacks in order to move to another location or to exit the warehouse. This means that the aisles have to be quite wide, using up warehouse space.

This example shows how synectics thinking can be used in the approach to such a problem. Direct analogies could be used to 'make the strange familiar', i.e. to familiarize the team with the new problem. For instance, analogies of the movement of snakes might be explored, leading to 'the problem as understood' being the need for a truck to twist sinuously in its manoeuvring. To 'make the familiar strange', the team might use personal and fantasy analogies of the kind: 'If I was holding the pallet in my outstretched arms, going along the aisle, I would like to be able to twist my upper body through ninety degrees (without moving my feet) to place the pallet in the rack.' Symbolic analogies of rotating turrets and articulated skeletons could lead eventually to a new design concept of an articulated truck with forks mounted on a front section that could swivel through ninety degrees. The Translift 'Benditruck' was designed on these principles.

Enlarging the search space

A common form of mental block to creative thinking is to assume rather narrow boundaries within which a solution is sought. Many creativity techniques are aids to enlarging the 'search space'.

Transformation

One such technique attempts to 'transform' the search for a solution from one area to another. This often involves applying verbs that will transform the problem in some way, such as

> magnify, minify, modify, unify, subdue, subtract, add, divide, multiply, repeat, replace, relax, dissolve, thicken, soften, harden, roughen, flatten, rotate, rearrange, reverse, combine, separate, substitute, eliminate.

Random input

Creativity can be triggered by random inputs from whatever source. This can be applied as a deliberate technique, e.g. opening a dictionary or other book and choosing a word at random and using that to stimulate thought on the problem in hand. Or switch on a television set and use the first visual image as the random input stimulus.

Why? Why? Why?

Another way of extending the search space is to ask a string of questions 'why?' about the problem, such as 'Why is this device necessary?' 'Why can't it be eliminated?', etc. Each answer is followed up, like a persistent child, with another 'Why?' until a dead end is reached or an unexpected answer prompts an idea for a solution. There may be several answers to any particular 'Why?', and these can be charted as a network of question-and-answer chains.

Counter-planning

This method is based on the concept of the dialectic, i.e. pitting an idea (the thesis) against its opposite (the antithesis) in order to generate a new idea (the synthesis). It can be used to challenge a conventional solution to a problem by proposing its deliberate opposite, and seeking a compromise. Alternatively, two completely different solutions can be deliberately generated, with the intention of combining the best features of each into a new synthesis.

The creative process

The methods above are some techniques which have been found useful when it is necessary for a designer or design team to 'turn on' their creative thinking. But creative, original ideas can also seem to occur quite spontaneously, without the use of any such aids to creative thinking. Is there, therefore, a more general process of creative thinking which can be developed?

Psychologists have studied accounts of creative thinking from a wide range of scientists, artists and designers. In fact, as most people have also experienced, these highly creative individuals generally report that they experience a very sudden creative insight that suggests a solution to the problem they have been working on. There is a sudden 'illumination', just like the light-bulb flashing on that cartoonists use to suggest someone having a bright idea.

This creative 'Ah-ha!' experience often occurs when the individual is not expecting it, and after a period when they have been thinking about something else. This is rather like the common phenomenon of suddenly remembering a name or word that could not be recalled when it was wanted.

However, the sudden illumination of a bright idea does not usually occur without considerable background work on a problem. The illumination or key insight is also usually just the germ of an idea that needs a lot of further work to develop it into a proper, complete solution to the problem.

Similar kinds of thought sequence occur often enough in creative thinking for psychologists to suggest that there is a general pattern to it. This general pattern is the following sequence: recognition – preparation – incubation – illumination – verification.

- *Recognition* is the first realization or acknowledgement that 'a problem' exists.

- *Preparation* is the application of deliberate effort to understand the problem.

- *Incubation* is a period of leaving it to 'mull over' in the mind, allowing one's subconscious to go to work.

- *Illumination* is the (often quite sudden) perception or formulation of the key idea.

- *Verification* is the hard work of developing and testing the idea.

This process is essentially one of work–relaxation–work, with the creative insight (if you are lucky enough to get one) occurring in a relaxation period. The hard work of preparation and verification is essential. Like most other kinds of creative activity, creative design is 1 % inspiration and 99 % perspiration!

The sudden illumination is often referred to as a 'creative leap', but it is perhaps not helpful to think of creative design as relying on a flying leap from the problem space into the solution space. The creative event in design is not so much a 'leap' from problem to solution as the building of a 'bridge' between the problem space and the solution space by the identification of a key solution concept. This concept is recognized by the designer as embodying a satisfactory match of relationships between problem and solution.

Rational Methods

More commonly regarded as 'design methods' than the creativity techniques are the rational methods which encourage a systematic approach to design. Nevertheless, these rational methods often have similar aims to the creative methods, such as widening the search space for potential solutions, or facilitating team work and group decision-making. So it is not necessarily true that rational methods are somehow the very opposite of creative methods.

Many designers are suspicious of rational methods, fearing that they are a 'straitjacket', or that they are stifling to creativity. This is a misunderstanding of the intentions of systematic design, which is meant to improve the quality of design decisions, and hence of the end product. Creative methods and rational methods are complementary aspects of a systematic approach to design. Rather than a straitjacket, they should be seen as a lifejacket, helping the designer, especially the student designer, to keep afloat.

Perhaps the simplest kind of rational method is the checklist. Everyone uses this method in daily life; for example, in the form of a shopping list, or list of things to remember to do. It *externalizes* what you have to do, so that you do not have to try to keep it all in your head, and so that you do not overlook something. It *formalizes* the process, by making a record of items which can be checked-off as they are collected or achieved until everything is complete. It also allows team work or participation by a wider group, e.g. all the family can contribute suggestions for the shopping list. It also allows subdivision of the task (i.e. improving the efficiency of the process), such as allocating separate sections of the list to different members of the team. In these respects, it is a model for most of the rational design methods. In design terms, a checklist may be a list of questions to be asked in the initial stages of design, or a list of features to be incorporated in the design, or a list of constraints, standards, etc., that the final design must meet.

There is a wide range of rational design methods, covering all aspects of the design process from problem clarification to detail design. The next eight chapters present a selection of the most relevant and widely used methods, also covering the whole design process. The selected set is listed below, with the stage in the design process on the left, and the method relevant to this stage on the right.

Identifying opportunities	User scenarios Aim: to identify and define an opportunity for a new or improved product.
Clarifying objectives	Objectives tree Aim: to clarify design objectives and subobjectives, and the relationships between them.
Establishing functions	Function analysis Aim: to establish the functions required, and the system boundary, of a new design.
Setting requirements	Performance specification Aim: to make an accurate specification of the performance required of a design solution.
Determining characteristics	Quality function deployment Aim: to set targets to be achieved for the engineering characteristics of a product, such that they satisfy customer requirements.
Generating alternatives	Morphological chart Aim: to generate the complete range of alternative design solutions for a product, and hence to widen the search for potential new solutions.

Evaluating alternatives

Weighted objectives
Aim: to compare the utility
values of alternative design
proposals, on the basis of
performance against
differentially weighted
objectives.

Improving details

Value engineering
Aim: to increase or maintain
the value of a product to its
purchaser whilst reducing its
cost to its producer.

As we shall discuss later, in Chapter 13, these eight stages of design,
and their accompanying design methods, should not be assumed
to constitute an invariate design process. However, Figure 4.1
suggests how they relate to each other and to the symmetrical
problem–solution model developed in Chapter 3. For example, clarifying objectives (using the objectives tree method) is appropriate
both for understanding the problem–solution relationship and for
developing from the overall problem into subproblems.

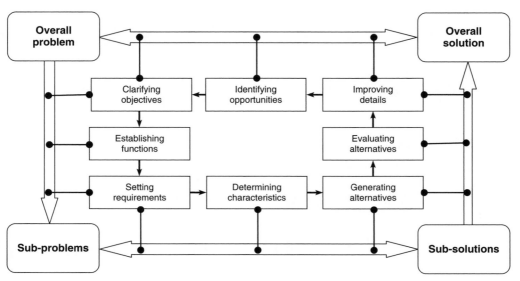

Figure 4.1 Eight stages of the design process positioned within the symmetrical problem/
solution model

This model of designing integrates the procedural aspects of design with the structural aspects of design problems. The procedural aspects are represented by the sequence of eight methods (anticlockwise, from the top), and the structural aspects are represented by the arrows showing the commutative relationship between problem and solution and the hierarchical relationships between problem/subproblems and between subsolutions/solution.

In the following eight chapters, each of the methods included in the model is presented in a step-by-step procedure, followed by a number of short practical examples and a more complete worked example. The examples show that such methods are often adapted to suit the particular requirements of the task in hand. Although it is important not to follow any method in a slavish and unimaginative fashion, it is also important that an effort is made to follow the principles of the method with some rigour. No beneficial results can be expected from slipshod attempts at 'method'.

5 Identifying Opportunities

Although most producer companies devote major resources to researching the market for their products, many products still appear on the market that seem not to have been designed with their user in mind. As modern products become more and more complex, with more and more functions and features available through their microprocessor components, many of them have become increasingly confusing to their users. At the same time, the availability of a greater diversity of products has led to greater consumer choice, and therefore the greater importance of consumer preferences. Consequently, a concern has grown to ensure that the design of products and systems becomes more user-centred.

Of course, there is not usually just one user of a product. Sometimes the range of users of a product, and their different needs, can be very diverse. In addition to the obvious or intended users there is a variety of people who have to interact with the product in various ways at different times, such as the people who make it, assemble it, service it and recycle it.

Many products do not work well in use because the presence of the user is not strongly represented during design. The designer is subject to many, often conflicting pressures that require tradeoffs to be made between users' needs and other factors. For example, manufacturing requirements can often override users' needs. This is because manufacturing and assembly difficulties are evident to the producer, and cost money through lost production time, faulty products, accidents and worker dissatisfaction.

A user-centred approach is often referred to as 'universal' or 'inclusive' design – meaning designing to include everyone.

Engineering Design Methods: Strategies for Product Design   N. Cross
© 2008 John Wiley & Sons, Ltd

Inclusive design means designing products so that they can be used easily by as many people as wish to do so. This may sound an obvious goal, but the fact is that many people have difficulty carrying out ordinary tasks with everyday products. Many elderly and disabled people cannot carry out – certainly with any ease or dignity – the range of everyday tasks that others take for granted. They are forced to choose products from a limited range that may suit them, and may also have to adapt products themselves, or buy expensive attachments to compensate for a product's inadequate design.

The *user scenarios* method provides a useful starting point and focus for the design process by taking the user's point of view, and identifying opportunities for creating new products that aim at satisfying users' needs.

The User Scenarios Method

Procedure

Practice being a user of a product or service

A simple first step in user research for design is to use a product or service yourself, but in an aware and critical frame of mind. This is called making a *user trip*, in which you use the product or service in a deliberate way, and note down your reactions.

The essential idea of user trips is very simple: you just take a 'trip' through the whole process of using a particular product or system, making yourself a critically observant user. You may be surprised how much you find out, if you make yourself sufficiently self-aware and observant. In fact, after trying this method a few times, you will probably find that you are adopting an attitude to most of the products and systems used in everyday life that other people might regard as hypercritical. But it is this critical attitude, this dissatisfaction with the accepted norms that spurs designers to make improvements. The secret is to turn dissatisfaction into creative criticism, to develop an attitude of constructive discontent.

Decide first of all which user's point of view you are taking: customer, operator, maintenance person, etc. You may want to make several trips, from different user perspectives, or try special user perspectives, such as that of a disabled person. It is usually easiest to take a customer's trip, since you may need special permission, access and perhaps skills before you can take any other.

Next, decide the limits, and the variations, to the user trip or trips that you are going to take. It is usually a good idea to extend the trip into activities that both precede and succeed the immediate use of the product you are investigating since this may lead you into devising an improved, more integrated overall solution. Similarly, variations on a basic trip (at different times of day, in different weather conditions, with different requirements) will probably bring to light a wider range of problems than would a single, random trip.

Now you just have to set out and take the trip for real, recording your actions, impressions, ideas and thoughts. On a user trip it is important to be on the lookout for even the very tiniest problems or discoveries, because they might add up to something significant. And do not ignore even your most trivial-seeming ideas for improvements, because again they might add up to something bigger, or lead later to an important insight into design faults. This recording of thoughts, impressions, etc., ideally should be done during the trip, or else as soon as possible afterwards. A pocket audio recorder is very useful; otherwise use a notebook.

If you are an experienced user of the product or system (using a familiar machine or following a regular procedure), remember that you will have adapted yourself to many of the difficulties involved, and will probably be unaware of the impressions that an inexperienced user would have. So try to make yourself again as naïve as you must once have been, and hypersensitive to even the slightest difficulties or impressions you may have. Alternatively select a situation or trip variation that makes you virtually an inexperienced user, such as using a new piece of equipment, or following an unfamiliar routine.

Observe users in action

User trips can generate a surprisingly large number of ideas for improvements to everyday products. But user trips suffer from a number of shortcomings that limit their usefulness. For example, a single product can have many different kinds of users and it is often difficult to take the viewpoint of a user who has a role different from your own. Or you might make mistakes that an experienced user would not, thus obtaining a wrong impression of the product. One way around these difficulties is to observe other people performing as users and to choose different kinds of experienced users, depending on what information you want to gather.

Begin by identifying those *experienced users* who will be able to provide you with relevant information. Most people are usually very willing to cooperate if approached tactfully and with obvious good intent. Get them to describe and demonstrate what seems important to *them*, as well as on what *you* regard as important. Look for, and ask about, aspects that seem difficult or unusual or critical. Look also for informal modifications to the job, workspace or equipment that users may have made for themselves (e.g. a handmade sign or label on a machine, additional loose cushions on a seat, etc.). Record what you see and are told, and your impressions, during or immediately after the observation.

An amazing amount of information relevant to the redesign of a product or system also can be gained from observing *inexperienced users'* attempts to cope. Inexperienced users can reveal potentially important problems, difficulties, ideas, etc., of which experienced users will be unaware.

Finding an inexperienced user is not always easy, but there may be ways of lowering a user's normal level of experience by introducing some novel feature. Acting as an inexperienced user, and perhaps being made to look naïve or foolish, can be embarrassing, too. You should reassure your volunteers (and maybe yourself), that you are looking for shortcomings in the product design, not in the users, and that they should blame the design of the product, not themselves, for any difficulties.

Once you have found volunteers, give them only the basic objective they have to achieve, without any detailed instructions. Such an objective might be 'Programme this DVD machine to record a programme on BBC 1 from 9.00 pm to 10.00 pm next Wednesday', or 'Listen to the voice-mail message that has been received on this telephone, and send a short text reply'.

You need to observe and record very carefully what the inexperienced user does. Get your volunteers to talk about the task as they attempt it. You may find that it takes what you would regard as an extraordinary amount of time for an inexperienced user successfully to operate something that you are familiar with yourself, and your volunteer may actually be unable to achieve a seemingly simple objective. Do not offer advice if your volunteers get stuck (unless it looks as though they will cause damage or an injury, or until it would be more fruitful to move on to the next stage of a sequence of operations), and try not to laugh or get angry. Remember the advice about blaming the design, not the user.

*Question users
about their
experiences*

All over the world, producer companies have increasingly learned to keep a careful watch on emerging consumer requirements and changing user needs and wishes. Techniques used in market research to gather consumers' views on products include both quantitative methods (collecting numerical data), such as survey questionnaires, and qualitative methods (collecting opinions and subjective responses), such as focus groups.

Most questionnaires require a fixed type of response, such as a choice between available answers, or along a scale of response. For example, a product design questionnaire might suggest 'I found the product easy to use' and provide a five-point scale of response from 'Agree strongly' to 'Disagree strongly'. Advantages of fixed-response types of questionnaires are that they are quick to complete, lend themselves to easy processing of the responses and result in numerical data.

Another type of questionnaire asks for open-ended responses, with questions such as 'Are there any features of the product that you dislike?' or 'Would you recommend this product to a friend – if so, why?' Open-ended questionnaires can provide useful insights gleaned from the responses, or can provide data such as how often a product feature is mentioned in free responses. But they require more processing of the responses, and provide more qualitative feedback and less numerical data than fixed-response questionnaires.

A 'focus group' is simply a group of people gathered together to discuss a particular issue. In research for product design, a focus group might be a group of purchasers of a particular product brought together to discuss their feelings and attitudes towards the product and rival products; or perhaps their general likes and dislikes about those types of products.

A focus group will typically have between six to a dozen participants, plus a discussion leader (the researcher). The leader will have a list of issues to guide the discussion, but will allow the participants to develop their own topics of discussion, so as to let the discussion go in directions favoured by the participants. Examples of particular products to be discussed will usually be available.

*Create relevant
user personas and
scenarios*

How can we draw together the varied needs and preferences, wishes and requirements into a 'model' of users of a new product, in order to provide some guidelines for the product designer? How can we use rather vague but interrelated sets of user preferences to identify the potential for new products to meet unsatisfied – and even unrecognized – user requirements? One way is to construct *scenarios* for how a product might be used by different users – that is, to construct some 'story lines' of the user's lifestyle and requirements that relate to opportunities for new product development.

Design scenario writing begins with identifying the potential *user profiles* and *personas*. A user profile summarizes the characteristics of the user population that must be designed for. A user persona is a hypothetical, but well-defined, particular user within that population. Personas are not real people; rather, they are imaginary examples of real users they represent.

In defining personas you should be as specific as possible about the 'made up' details, and you should also give the persona a name, and possibly an image, whether it is an anonymous photograph or a sketched caricature. During the design process, the persona is referred to by name, rather than as 'the user'. A set of personas can be defined for a project, something like a 'cast of characters' who might use the product. Every cast, though, should have at least one primary persona who is the main focus of your design.

Observing and talking to users can provide starting points for identifying profiles and personas. For instance, you can observe the range of different users of a particular product. Or if you know some people who are likely to be users of a certain product type, you can draw upon your personal knowledge of these people, or you can ask them to complete a questionnaire about their lifestyle, likes and dislikes. Different product types will have different user profiles.

Define the preliminary goal, context, constraints and criteria for a new product opportunity

The formal starting point for the design of a new product is the design brief. Although it is a very important step in the product planning and development process, it does not always get the attention it deserves. A design brief is the instruction to the designer from the client to take on a project, and some background information on the decision to initiate the project. However, in many examples of product design, this instruction is 'brief' in another sense – i.e. it contains very little information!

The design brief is an important step in the clarification of just what a new product is expected to be and to do. It is not, of course, a statement of the design *solution* – which is what the designer(s) will eventually generate – but a statement of the design *problem* or *opportunity*.

The design brief is an intermediate stage between the funda-mental or underlying product idea and a full specification of the requirements that the product is expected to satisfy. (Developing a full specification is dealt with in the performance specification method, Chapter 8.) Crucially, the brief sets the design *goal*, outlines the *context* in which the new product must operate, identifies the major *constraints* within which the design goal must be achieved and suggests some *criteria* by which a good design proposal might be recognized. The difference between constraints and criteria is that constraints set specific (usually quantitative) targets or limits for the product designer, whilst criteria are more flexible and might be used for judging between different product design proposals, each of which meets the specific constraint targets.

Summary

The aim of the user scenarios method is to identify and define an opportunity for a new or improved product. The procedure is as follows:

1. Practice being a user of a product or service. Decide which user's (or users') point(s) of view to adopt and the variations to the user trip or trips you are going to take.

2. Observe users in action. Both experienced and inexperienced users can provide valuable insights.

3. Question users about their experiences. This can include the use of formal, structured or unstructured questionnaires, and focused group discussions.

4. Create relevant user personas and scenarios. A persona is a well-defined but hypothetical user, and a scenario is a storyline about their use of the product or service.

5. Define the preliminary goal, context, constraints and criteria for a new product opportunity. These are the key steps in formulating a good brief for a new product design.

Examples

Example 1: sewing machine

A manufacturer of domestic sewing machines wanted to revamp one of its products. The manufacturer's aim was to keep the product looking fresh, in order to maintain its market position. The manufac-turer asked a designer to suggest a new design, perhaps thinking only of some restyling. The resulting new product design incorpo-rated the manufacturer's standard machinery, but repackaged it in novel ways that not only made it easier to use but also gave the machine a new and distinctive form and style.

The origins of the new design features lay in the designer's personal experience through a user trip with the existing model of sewing machine, using it for some simple sewing tasks. The designer found it strange that the sewing arm was located centrally over its base, whereas he realised that the user needs more surface area on their side of the needle than behind it. In front of the needle, space is needed to prepare the material for sewing; once the work has passed beyond the needle the user can do nothing more: it is sewn. The designer therefore proposed moving the sewing machine mechanism rearwards on its base, creating an off-centre layout with more base-table area in front of the needle than behind it. Traditionally, the sewing machine had been designed as a very straightforward, basic piece of engineering with the mechanism placed centrally upon the base, and nobody had thought about challenging this.

Another radical change in this particular design was also a result of a simple assessment of how the machine is used. The designer gave the base of the machine rounded lower edges, which look like a mere 'styling' feature, but in fact also arose from considering use. A frequent problem with sewing machines is that a lot of lint and loose fibres come off the fabric. These get down into the bobbin and shuttle mechanism below the needle, may get picked up and sewn into the workpiece and at worst may stop the machine altogether. The traditional way of dealing with this was to require the user to open the base to get at the mechanism for cleaning. This was usually achieved by tilting the machine away from the user, into an unstable position that only allowed restricted access to the mechanism.

The designer decided to make access to the underneath much easier, by tilting the machine lengthways. That action in itself suggested a shape by giving the lower side edge of the machine a rounded shape to make it easier for the user to tilt the machine, and the underside was more easily accessible for cleaning and oiling the lower mechanisms. A similar radiused top front edge was also provided to the base plate, to allow the fabric to slide over it more easily, and various other features were added, such as small drawers for holding accessories. The designer's sketches (Figure 5.1) help explain these ideas and features. (Source: Grange, 1999.)

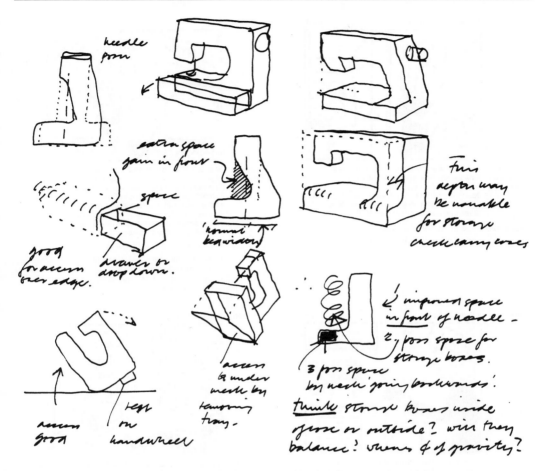

Figure 5.1 Design sketches for the user-centred redesign of a sewing machine

Example 2: digital radio

When a broadcasting company wanted to explore the potential of digital audio broadcasting, it asked a design consultancy to suggest new digital radio product concepts. The team used personas and scenario building to help generate ideas for the functions, operation and appearance of new radios. The ideas included visual displays built in to the radios to complement the audio programme, such as musician portraits, programme guides, etc. One project focused on special purpose radios used by one person in different situations. Figure 5.2 shows one user persona that the designers created – Jean, whose work sometimes requires her to be on 24-hour call. The comments attached to the scenarios include ideas for features to be incorporated in a new radio design, prompted by the scenarios. (Source: Myerson/IDEO, 2001.)

scenario **1 Jean's emergency**   2.00 am

scenario description   Jean is woken by a telephone call, it's an emergency. She gets out of bed and starts to get ready.
Jean switches on her personal digital radio in the bathroom to help her wake up. It automatically tunes into her usual channel for this time of the day.

key ideas   • specific stations for specific time – 'MYradio'

      • agent driven learning system – watch me

scenario **2 In the car**   2.20 am

scenario description   She gets into her car and docks her personal digital radio to recharge the batteries.
The radio communicates her car-specific listening preferences, via infra-red (IR) link, to the in-car radio, so when she wants to listen to music the car radio suggests her favourite station. Her programme is interrupted by some travel news but luckily it will not affect her journey.

key ideas   • car docking – powers battery

      • infra-red link – personal radio can communicate preferences to other radios

      • environment specific information – in the car she gets travel news

scenario **3 Back at home**   5–6.00 am

scenario description   Jean arrives back home. As she moves from room to room her digital radio automatically links to the speakers in each room. She is listening to BBC Radio Music Plus as usual for this time in the morning.
Knowing she wants some music to go jogging to later, Jean records some music from the Fitness Channel. She stores it on a memory disc as she may want to use it again.
The programme has already started but she is able to record it from the beginning.

key ideas   • speakers with infra-red connection – auto, manual settings

      • audio download to storage

      • recording from the start of programme

scenario **4 In her lounge**   12.05 pm

scenario description   After her run, Jean rests in the lounge and browses the latest copy of Digital Radio Times. This magazine is downloaded once a week and gives Jean scheduling information as well as interesting articles. She sets her digital radio to view through her projector so she can see the text easily.
As she views the schedule information, different programmes are subtly highlighted due to her preference and previous listening habits. She scans the information both visually and audibly.

key ideas   • visual manipulation of information to show preference

      • audio visual magazine

      • bulk download – probably at night

      • projected displays

Figure 5.2   Scenarios for digital radio use by a shift worker

Example 3: oven aid

In this example, observation of elderly people using equipment in their kitchens led to identification of a new product opportunity. The designers developed a persona, 'Mary', and identified a particular problem that she (and millions like her) has in using her oven. The example develops a new product opportunity from the persona and a user scenario. (Source: Cagan and Vogel, 2002.)

Background

Many elderly people experience difficulty with lifting things into and out of their ovens. This spoils their enjoyment of an everyday task, cooking, and restricts their independence in being able to perform such a task. The difficulties can be due to a variety of physical factors such as arthritis and rheumatism in hands, arms and back, muscular weakness and limitations in movements such as bending and reaching.

The primary product users in this context were considered to be elderly women. Although the problem is applicable to men as well, the majority of the elderly population is female. Any particular issues for women may make the product better meet the majority needs, wants and desires of this population.

As well as observing elderly women performing cooking tasks with their ovens in their own homes, the design team consulted expert advisers such as health care workers who work with the elderly and doctors who work with seniors and are experts in rheumatology. They also researched guidelines for designing for disability. Other stakeholders consulted included people who install ovens and sell appliances, and organizations that promote products for seniors. It was also important to know about ovens, especially the type of ovens that older women might own.

Persona

Mary is 70 years old and lives alone, but loves to bake and to entertain her family at home for holidays. Mary has arthritis in the lower spine and shoulders that limits her range of motion. She also has lost strength in her back and arm muscles. She is no longer comfortable reaching into the oven to lift things out. Losing the ability to bake things has been very depressing for her to contemplate. Mary is becoming hesitant to have her family over and no longer feels confident entertaining in her home.

User scenario

Mary needs help in lifting relatively heavy dishes into and out of her oven. A device is needed that fits with a standard oven and will compensate for Mary's limited motion and reduced strength and allow her to easily put in and remove from her oven a variety of pans and baking dishes. The device must have a simple, robust operating mechanism and must be inexpensive to buy and install. While the primary user market is senior women between the ages of 70 and 85, the primary purchasers may be family members. Any installation should be easy enough for a family member to do. The product must integrate with a range of standard ovens and be easy to clean.

*Product
opportunity
proposal*

This product opportunity can be summarized in terms of the design goal, context, constraints and criteria as follows:

- *Goal:* an aid to assist lifting pots and dishes into and out of domestic ovens.

- *Context:* elderly people, especially women, who have problems lifting heavy dishes into and out of their ovens in everyday cooking.

- *Constraints:*

  - usable by most women aged 70–85

  - selling price not more than £30

  - can lift and support a range of cooking pots, baking trays, etc., and their contents

  - integrates with standard conventional ovens

  - safe

  - simple installation.

- *Criteria:*

  - attractive to purchasers as well as users

  - easy to use

  - appears robust and safe.

*Worked example:
message recorder*

An electronics communications company appointed a design consultant to identify areas for new product opportunities. The company wanted suggestions for design briefs that could be handed to their own product designers for development into specific product concepts.

Here is an outline of how the consultant approached one area: that of telephone answering machines. She realized that telephone answering machines are a product type that may have begun to outlive its natural life, as many users have taken up greater use of cell phones with text and voice-mail facilities. However, there may still be opportunities for products of this type, aimed at specific consumer markets, such as older people.

The consultant realized that she and her family have two ageing answering machines in their house, but they often forget to set them, or forget to check for messages. Also, she seems to be the only person who can operate them, being the only one who has read the manuals. Their first machine was bought about 12 years previously, and is integral with a telephone handset. The second machine was bought about five years ago, and is integral with a fax machine. Both machines seem to be confusing to inexperienced users, and inadequate for normal family use, so she thinks there is scope for improvement and possible new product development.

The consultant first tried observing other users. She asked a variety of family members and friends to use the two answering machines. She set the same set of tasks for each machine, which were to:

- listen to the previously received message that she had telephoned into both machines;

- act on the message, which was to note or remember a telephone number and pass it on to someone else;

- save the message;

- record their own answer 'announcement', which will be heard when anyone phones in and gets the answer machine;

- set the machine to receive messages.

The consultant asked these trial users to think aloud as they were doing these tasks, and recorded what they said on a small audio recorder. She also made her own notes as each trial user performed the tasks. For example, she noted that most users found it difficult to memorize the telephone number that the caller left in the message, and there was no pencil and paper at hand to write it down. From this observation, she noted the idea that an answering machine might be able to provide a printout of callers' numbers, with date and time of the call.

The consultant went on to conduct some focus group research. She asked small groups of target group users, aged in their fifties and sixties, to discuss answering machines together, using pairs of various machines as examples to initiate the discussion. As the group facilitator, she had prepared several prompts to keep the discussion going, and to move it into areas that she thought might be interesting. Some of the prompts were:

- What differences do you see between the two machines?

- Does one look easier to use? Why?

- What are the poor aspects of answering machines that you can recall from your own personal experiences?

- How do you feel about finding that your call is being taken by an answering machine?

- Can you recall either a funny event or a serious problem arising from missing a telephone message or misunderstanding a message?

- Besides telephone messages, what other messages do you leave around the house? Are there messages you would like to leave, but cannot?

One especially appropriate idea to emerge from the focus groups was a suggestion that an answering machine might be a more general 'message machine' that would allow anyone to record a message on it (like an audio recorder) for anyone else in the household who comes home later – for example, 'I've gone to mother' or 'Your dinner's in the oven'.

User profile

The consultant proposed that the market segment for new message recording machines is people aged 55 and older, who perhaps have no mobile telephone or only use one infrequently. Table 5.1 is the user profile she developed.

The consultant developed the following persona and scenario for someone falling within the user profile.

User persona

Brian is aged 62 and has recently retired as an accounts manager in a transport company. He has always been quite active, playing squash at weekends or going swimming in his lunch hour. Now he is retired, he has taken up some additional activities, including joining a sailing club, and spends more time on activities such as gardening. He is often away from home, or outside the house, for two to three hours at a time, most days of the week. He and his wife, Brenda, have also started taking occasional long weekend breaks away from home. They have active social lives, both separately and together, with large networks of friends and family.

Table 5.1 User profile for a message recorder.

User characteristic	User profile
Age	55–75
Sex	Male and female
Physical limitations	May have some relevant physical limitations, such as impaired manual dexterity, arthritis or reduced acuity of vision and hearing
Telephone and message recording experience	Could have quite limited previous experience; not a frequent mobile-phone user; unfamiliar with voice-mail systems
Motivation	Wants to be sure not to miss incoming messages
Attitude	Exasperated by telephones and other machines that have difficult control buttons, poor legibility of graphics and complicated procedures of use

User scenario

Previously, at work, Brian always had a telephone on his desk, and could make and receive personal calls as well as his work-related calls. If he was away from his desk, someone else in the office would pick up his calls. He does not have an answering machine at home, because it has not seemed necessary. Now, he knows that he often misses calls that friends and family make. They complain that it is now difficult to contact him, whereas it was a lot easier when he was at work. Brian does not have a mobile phone, and has a fairly negative attitude towards them. His wife Brenda does have a mobile phone, which was given to her as a Christmas present two years ago by their son. She regards it as for 'emergency use', such as when they are away from home, or travelling in the car, and rarely has it switched on. She finds its small buttons difficult to read and fiddly to use. Both Brian and Brenda could make use of a domestic telephone answering machine that was easy to use and also allowed them to leave messages for each other.

Product opportunity

The consultant framed a proposal for a new product opportunity – a message recorder – that would suit the Brian persona, as follows.

- Although voice-mail and text messaging systems have grown in use, this has mainly been among younger users of mobile telephones. There is also a growing market among older users of landline telephones for improved and new versions of telephone answering and other message recording machines. There is a new product opportunity for a stand-alone, domestic telephone answering machine/message recorder for this market segment.

- Potential users are aged 55–75, having various previous experiences of telephone systems, from a little to a lot. They are not regular users of mobile telephones. They have active lives and are often out of the house, but do not want to miss incoming messages from friends and family.

- For this type of user, the machine must be simple to use, and appear to be so at the point of purchase. This implies easy-to-read graphics and clear, easy-to-use controls. Setting the machine to record messages and the playing back of messages must be simple, obvious operations.

- Total recording time for messages should be at least 15 minutes. There should be a facility to use the machine as an audio recorder for messages left for each other between members of the household. There should be a highly visible alert signal that shows a message has been received. There should be a facility to print out the originating number, and date and time of each telephone message.

Summary proposal

- *Goal:* a stand-alone, domestic telephone and other message recording machine.

- *Context:* potential users are middle-aged or elderly, perhaps not mobile phone users, anxious not to miss incoming messages when they are away from home.

- *Constraints:*

  - usable by people with mildly impaired dexterity, vision, hearing

  - easy to set record function and to play back messages

  - total recording time at least 15 minutes

  - highly visible alert signal for message waiting

  - printout facility for message originating number, date, time

  – microphone facility to record a message on the machine

  – retail selling price under £20.

- *Criteria:*

  – easy to use

  – user-friendly appearance.

6 Clarifying Objectives

When a client, sponsor or company manager first approaches a designer with a product need, it is unlikely that the 'need' will be expressed as clearly as in the previous chapter. The client perhaps knows only the type of product that is wanted, and has little idea of the details, or of the variants that might be possible. Or the 'need' might be much vaguer still: simply a 'problem' that needs a solution.

The starting point for design work is therefore very often an ill-defined problem, or a rather vague requirement. It will be quite rare for a designer to be given a complete and clear statement of design objectives. Yet the designer must have some objectives to work towards. The outcome of designing is a proposal for some means to achieve a desired end. That 'end' is the set of objectives that the designed object should meet.

An important first step in designing therefore is to try to clarify the design objectives. In fact, it is very helpful at all stages of designing to have a clear idea of the objectives, even though those objectives may change as the design work progresses. The initial and interim objectives may change, expand or contract, or be completely altered as the problem becomes better understood and as solution ideas develop.

So it is quite likely that both 'ends' and 'means' will change during the design process. But as an aid to controlling and managing the design process it is important to have, at all times, a statement of

Engineering Design Methods: Strategies for Product Design   N. Cross
© 2008 John Wiley & Sons, Ltd

objectives which is as clear as possible. This statement should be in a form which is easily understood and which can be agreed by the client and the designer, or by the various members of the design team. (It is surprising how often members of the same team can have different objectives!)

The *objectives tree* method offers a clear and useful format for such a statement of objectives. It shows the objectives and the general means for achieving them which are under consideration. It shows in a diagrammatic form the ways in which different objectives are related to each other, and the hierarchical pattern of objectives and subobjectives. The procedure for arriving at an objectives tree helps to clarify the objectives and to reach agreement between clients, managers and members of the design team.

The Objectives Tree Method

Procedure

Prepare a list of design objectives

The 'brief' for a design problem is often very aptly called that – it is a very brief statement! Such brevity may be because the client is very uncertain about what is wanted, or it may be because he or she assumes that the designer perfectly understands what is wanted. Another alternative is that the client wishes to leave the designer with as much freedom as possible. This might sound like a distinct advantage to the designer, but can lead to great frustration when the client decides that the final design proposal is definitely *not* what was wanted! In any case, the designer will almost certainly need to develop the initial brief into a clear statement of design objectives.

The design objectives might also be called client requirements, user needs or product purpose. Whatever they are called, they are the mixture of abstract and concrete aims that the design must try to satisfy or achieve.

Some design objectives will be contained within the design brief; others must be obtained by questioning the client, or by discussion in the design team. Typically, initial statements of objectives will be brief and rather vague, such as 'The product must be safe and reliable'. To produce more precise objectives, you will need to expand and to clarify such statements.

One way to begin to make vague statements more specific is, literally, to try to *specify* what it means. Ask 'what is *meant* by that statement?' For example, an objective for a machine tool that it must be 'safe', might be expanded to mean:

1. Low risk of injury to operator.

2. Low risk of operator mistakes.

3. Low risk of damage to workpiece or tool.

4. Automatic cut-out on overload.

This kind of list can be generated simply at random as you think about the objective, or in discussion within the design team. The client may also have to be asked to be more specific about objectives included in the design brief.

The types of questions that are useful in expanding and clarifying objectives are the simple ones of 'why?', 'how?' and 'what?' For instance, ask 'why do we want to achieve this objective?', 'how can we achieve it?' and 'what implicit objectives underlie the stated ones?' or 'what is the problem really about?'

Order the list into sets of higher-level and lower-level objectives   As you expand the list of objectives it should become clear that some are at higher levels of importance than others. Subobjectives for meeting higher-level objectives may also emerge, and some of the statements will be means of achieving certain objectives. This is because some of the questions that you will have been asking about the general objectives imply a 'means–end' relationship – that is, a lower-level objective is a means to achieving a higher-level one.

An example is the statement 'automatic cut-out on overload' in the list above. This is not really an objective in itself, but a means of achieving an objective – in this case, the objective of 'low risk of damage to workpiece or tool'. In turn, this 'low-risk of damage' objective is itself a lower-level objective to that of the overall 'safety' objective.

Your expanded list of objectives will therefore inevitably contain statements at various levels of specificity. In order to clarify the various levels that are emerging, rewrite your general list of objectives into ordered sets. That is, group the objectives into sets, each concerned with one highest-level objective. For example, one set might be to do with 'safety', another to do with 'reliability', and so on. Within each set, list the subobjectives in hierarchical order, so that the lower-level ones are clearly separated as means of achieving the higher-level ones. Thus, for instance, your 'safety' list might look like this:

- Machine must be safe

- Low risk of injury to operator
- Low risk of operator mistakes
- Low risk of damage to workpiece or tool

- Automatic cut-out on overload

The list is now ordered into three hierarchical levels. It can sometimes be difficult to differentiate between levels of objectives, or different people in the design team may disagree about relative levels of importance of some objectives. However, exact precision of relative levels is not important, and you want only a few levels, about which most people can agree. For instance, in the above list, 'low risk of injury' might be considered more important than 'low risk of mistakes', but all three 'low risk' objectives can conveniently be grouped at about the same level.

The valuable aspect to sorting objectives roughly into levels is that it encourages you to think more clearly about the objectives, and about the relationships between means and ends. As you write out your lists in hierarchical levels, you will probably also continue to expand them, as you think of further means to meet subobjectives to meet objectives, etc.

When you have quite a lot of statements of objectives, it is easier to sort them into ordered sets if each statement is written onto a separate slip of paper or small card. Then you can more easily shuffle them about into groups and levels.

Draw a diagrammatic tree of objectives, showing hierarchical relationships and interconnections

As you write out and shuffle your lists, you will probably realize that some of the subobjectives relate to, or are means of achieving, more than one higher-level objective. For example, the subobjective of 'low risk of damage to workpiece or tool' might be not only a means of achieving safety but also a means of achieving reliability.

So a diagram of the hierarchical relationships of these few objectives and subobjectives might look like that in Figure 6.1. This diagram is the beginnings of a 'tree' which shows the full pattern of relationships and interconnections. It is not necessarily just a simple 'tree' structure of branches, twigs and leaves, because some

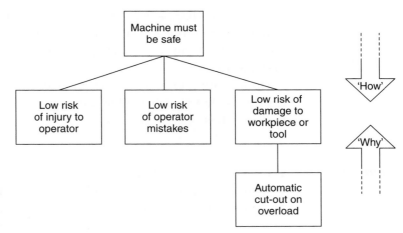

Figure 6.1
Hierarchical
diagram of
relationships
between objectives

of the interconnections form loops or lattices. The 'tree' is also normally drawn 'upside-down' – i.e. usually it has increasingly more 'branches' at lower levels – and so it might be better to think of the subobjectives as 'roots' rather than 'branches'.

It can sometimes be more convenient to draw the tree on its side, i.e. with branches or roots spreading horizontally. In order to help organize the relationships and interconnections between objectives and subobjectives, draw a complete tree diagram, based on your ordered sets of objectives. Each connecting link that you draw indicates that a lower-level objective is *a means of achieving* the higher-level objective to which it is linked. Therefore working *down* the tree a link indicates *how* a higher-level objective might be achieved; working *up* the tree a link indicates *why* a lower-level objective is included.

Different people might well draw different objectives trees for the same problem, or even from the same set of objectives statements. The tree diagram simply represents one perception of the problem structure. The tree diagram helps to sharpen and improve your own perception of the problem, or to reach consensus about objectives in a team. It is also only a temporary pattern, which will probably change as the design process proceeds.

As with many other design methods, it is not so much the end product of the method (in this case, the tree diagram) which is itself of most value, but the process of working through the method. The objectives tree method forces you to ask questions about objectives, such as 'What does the client mean by X?' Such questions help to make the design objectives more explicit, and bring them into the open for discussion. Writing the lists and drawing the tree also begin the process of suggesting means of achieving the design objectives, and thus of beginning the process of devising potential design solutions.

Throughout a project, the design objectives should be stated as clearly as the available information permits; the objectives tree facilitates this.

Summary

The aim of the objectives tree method is to clarify design objectives and subobjectives, and the relationships between them. The procedure is as follows:

1. Prepare a list of design objectives. These are taken from the design brief, from questions to the client and from discussion in the design team.

2. Order the list into sets of higher-level and lower-level objectives. The expanded list of objectives and subobjectives is grouped roughly into hierarchical levels.

3. Draw a diagrammatic tree of objectives, showing hierarchical relationships and interconnections. The branches (or roots) in the tree represent relationships which suggest means of achieving objectives.

Examples

Example 1: city transport system

This is an example of expanding and clarifying design objectives from an initially vague brief. A city planning authority asked a transport design team for proposals for 'a modern system, such as a monorail, which would prevent traffic congestion in the city from getting any worse and preferably remove it altogether'.

The only clear objective in this statement is 'to prevent traffic congestion . . . from getting any worse . . . ' But what are the implicit objectives behind the desire for 'a modern system, such as a monorail'? Traffic congestion might be held constant or reduced by other means.

By questioning the client, the design team uncovered objectives such as a desire to generate prestige for the city and to reflect a progressive image for the city authority. There was also a wish simply to reduce complaints from citizens about the existing traffic system. It was also discovered that only certain types of new system would be eligible for a subsidy from central government.

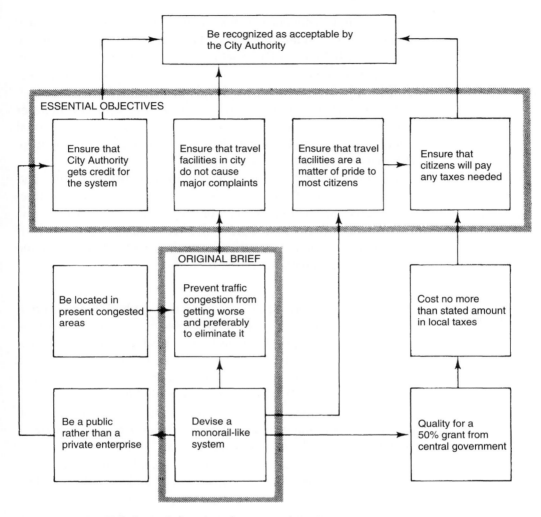

Figure 6.2 Expanded set of objectives for a new city transport system

The design team were able to draw up an expanded and hierarchically ordered set of objectives, as shown in Figure 6.2. In particular, they identified a number of high-level, 'essential objectives' which were not explicitly stated in the original brief. By identifying these objectives, the designers clarified the project and the limitations that there might be on the range of alternative solutions. (Source: Jones, 1992.)

Example 2: regional transport system

Another example from transport design is shown here, for a larger, regional system. The designers started from the clients' vague definition of 'a convenient, safe, attractive system', and expanded each objective in turn.

For example 'convenience' was defined in terms of 'low journey times' and 'low "out-of-pocket" costs' for users. The latter objective can be met by appropriate pricing policies; low journey times can be met by a variety of subobjectives, as shown on the left-hand side of the objectives tree in Figure 6.3.

Two aspects of 'attractiveness' were defined: user and nonuser aspects. The user aspects were subdivided into comfort, visual appeal and internal noise, whereas the nonuser aspects were external noise and visual obtrusiveness.

The 'safety' objective was defined to include deaths, injuries and property damage. The subobjectives to these show how subobjectives can contribute to more than one higher-level objective. A 'low risk of accidents' can contribute to all three higher-level objectives. If accidents do occur, a 'low risk of injury per accident' can contribute to keeping down both injuries and deaths.

Example 3: impulse-loading test rig

An example of applying the objectives tree method in engineering design is provided here. The design problem was that of a machine to be used in testing shaft connections subjected to impulse loads.

As before, a typically vague requirement of a 'reliable and simple testing device' can be expanded into a much more detailed set of objectives (Figure 6.4). 'Reliability' is expanded into 'reliable operation' and 'high safety'. 'Simple' is expanded into 'simple production' and 'good operating characteristics'; the latter is further defined as 'easy maintenance' and 'easy handling'; and so on.

In a case such as this, first attempts at expanding the list of objectives would probably produce statements at all levels of generality. For example, asking 'What is meant by "simple"?' would have been likely to produce statements in random order such as, 'easy maintenance', 'small number of components', 'simple assembly', etc. Drawing these out in the hierarchical tree structure shows how they relate together. (Source: Pahl and Beitz, 1999.)

Example 4: automatic teamaker

The objectives tree method can also be used in designing a relatively simple device such as an automatic teamaker. In this example, a distinction is made between 'functions' and 'means'. Each 'function' is an objective, which may be achieved by a number of different 'means' or subobjectives. Thus the function 'combine water and tea leaves' could be achieved by adding the water to the tea, adding the tea to the water, or bringing them both together into one receptacle (Figure 6.5).

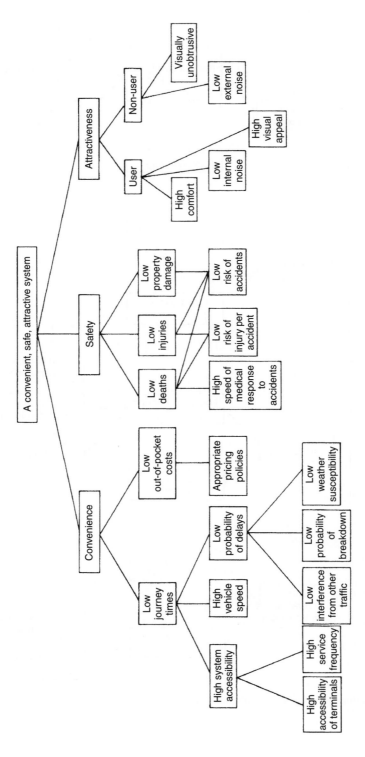

Figure 6.3 Objectives tree for a 'convenient, safe, attractive' new transport system

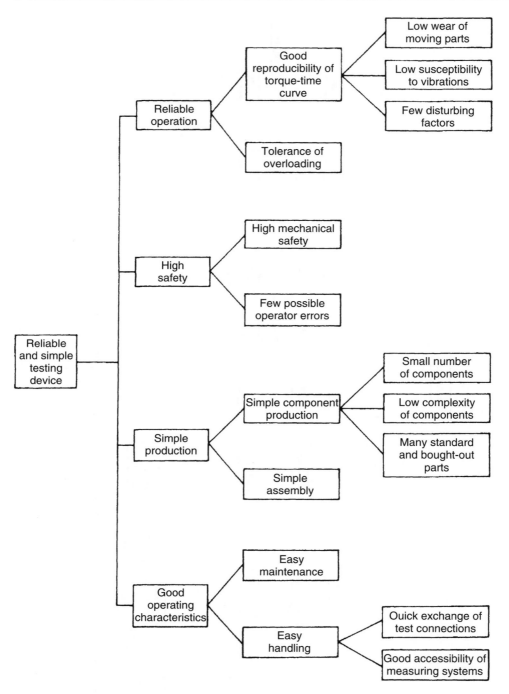

Figure 6.4   Objectives tree for an impulse-loading test rig

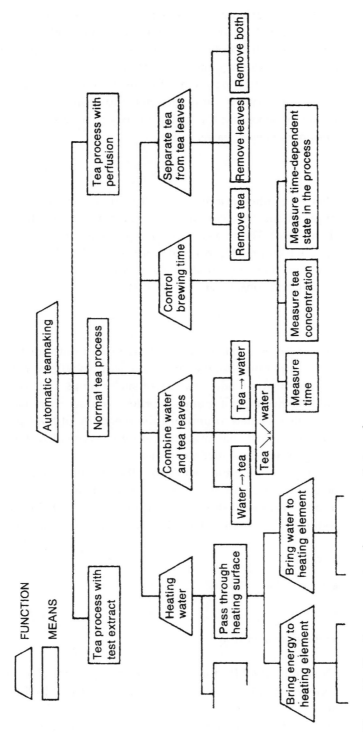

Figure 6.5   Functions–means tree for an automatic teamaker

This is a variation on the objectives tree as described earlier and demonstrated in the other examples, and might more accurately be called a 'functions tree'. However, the same principles apply of breaking down objectives into subobjectives, or functions into means, and ordering them into a hierarchical tree. This application of the tree structure approach helps to ensure that all the possible means of achieving a function (or objective) are considered by the designer. (Source: Tjalve, 1979.)

Example 5: car door

This is another example of a 'function tree'. In considering the requirements for a car door, the designers set out a tree of functions (Figure 6.6). The tree starts from high-level functional requirements (on the left of the figure), and works through to lower-level, detailed functions that can actually be implemented in terms of engineering design decisions. (Source: Pugh, 1991.)

Worked example: high-pressure pump

This example is based on the design of a pump for high-pressure, high-temperature fluids. The manufacturers who commissioned the design already made a variety of such pumps, but wished to rationalize their range of pumps in order to reduce manufacturing costs. They also wanted to improve the reliability of their pumps and to offer a product that was seen to be convenient to the varied and changing needs of their customers.

On questioning the client about the 'reliability' and 'convenience' objectives, a common aspect emerged: that the pump should be 'robust', i.e. that it should not easily fail. The initial list of objectives so far might therefore look like this, in hierarchical order:

- Reliable

- Convenient

- Robust

- Standardized range.

These are all still rather high-level and general objectives, so it is necessary to investigate such statements further. In this case, it was possible to investigate the problems experienced with the existing pumps. It was found that they were sometimes affected by cracking

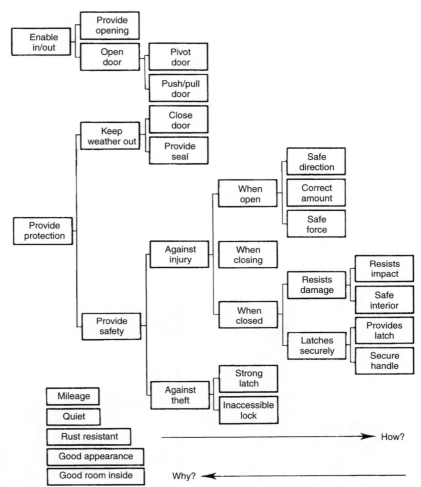

Figure 6.6
Function tree for a
car door

and leakages due to the stresses caused by the thermal expansion of the pipes to which they were connected. This appeared to be the main problem to which the requests for 'robustness' and 'reliability' were aimed.

Similarly, investigating the 'convenient' objective revealed a further two subobjectives: firstly, that the pumps should be easy to install and replace; and secondly, that they should occupy the minimum space. It was realized that the standardization of sizes and dimensions in the range could be a means of helping to achieve these objectives, as well as reducing manufacturing costs.

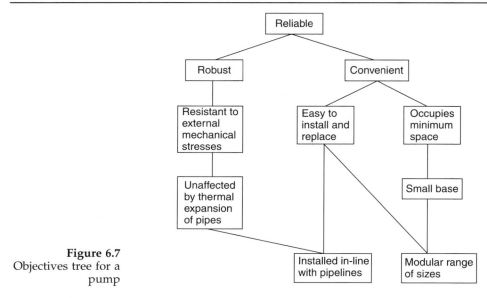

Figure 6.7
Objectives tree for a
pump

Figure 6.8 The Grundfos pumps that were designed on similar principles to the objectives developed in this example

The expanded list of objectives therefore looked like this:

- Reliable

- Robust

- Resistant to external
  mechanical stresses

- Unaffected by thermal
  expansion of pipes

- Convenient

- Easy to install and replace

- Occupy minimum space

- Standardized range

A key design principle to emerge from considering the means of achieving these objectives was that the inlet and outlet ports should always be in-line, to avoid the thermal expansion problems. Such a system, coupled with a small base size and modular dimensioning of alternative components, would also facilitate installation and replacement of the pump. The objectives tree therefore looked like that in Figure 6.7.

A high-pressure pump has been designed on similar principles in Denmark (see Figure 6.8). According to a description by the Danish Design Council, the pump is 'almost a diagram of its problem statement: intake and discharge are aligned, motor, coupling and the stage-built pump are aligned on an axis at right angles to the installation surface, and the pump pressure is increased by adding to the number of stages, i.e. a change in height. The pump is installed directly on the pipeline, occupying a minimum of space.'

7 *Establishing Functions*

We have seen from the objectives tree method that design problems can have many different levels of generality or detail. Obviously, the level at which the problem is defined for or by the designer is crucial. There is a big difference between being asked to 'design a telephone handset' or to 'design a telecommunication system'.

It is always possible to move up or down the levels of generality in a design problem. The classic case is that of the problem 'to design a doorknob'. The designer can move up several levels to that of designing the door or even to designing 'a means of ingress and egress' and find solutions which need no doorknob at all – but this is of no use to a client who manufactures doorknobs! Alternatively, the designer can move down several levels, investigating the ergonomics of handles or the kinematics of latch mechanisms – perhaps again producing non-doorknob solutions which are functional improvements but which are not what the client wanted.

However, there are often occasions when it is appropriate to question the level at which a design problem is posed. A client may be focusing too narrowly on a certain level of problem definition, when a resolution at another level might be better, and reconsidering the level of problem definition is often a stimulus to the designer to propose more radical or innovative types of solutions.

So it is useful to have a means of considering the problem level at which a designer or design team is to work. It is also very useful if this can be done in a way that considers not the potential type of solution, but the essential functions that a solution type will be required to satisfy. This leaves the designer free to develop alternative solution proposals that satisfy the functional requirements.

The *function analysis* method offers such a means of considering essential functions and the level at which the problem is to be addressed. The essential functions are those that the device, product or system to be designed must satisfy, no matter what physical components might be used. The problem level is decided by establishing a 'boundary' around a coherent subset of functions.

The Function Analysis Method

Procedure

Express the overall function for the design in terms of the conversion of inputs into outputs

The starting point for this method is to concentrate on what has to be achieved by a new design, and not on how it is to be achieved. The simplest and most basic way of expressing this is to represent the product or device to be designed as simply a 'black box' which converts certain 'inputs' into desired 'outputs'. The black box contains all the functions that are necessary for converting the inputs into the outputs (Figure 7.1).

It is preferable to try to make this overall function as broad as possible at first – it can be narrowed down later if necessary. It would be wrong to start with an unnecessarily limited overall function which restricts the range of possible solutions. The designer can make a distinct contribution to this stage of the design process by asking the clients or users for definitions of the fundamental purpose of the product or device, and asking about the required inputs and outputs: from where do the inputs come, what are the outputs for, what is the next stage of conversion?, etc.

This kind of questioning is known as 'widening the system boundary'. The 'system boundary' is the conceptual boundary that is used to define the function of the product or device. Often, this boundary is defined too narrowly, with the result that only minor design changes can be made, rather than a radical rethinking.

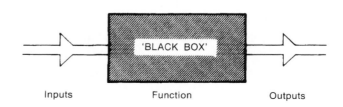

Figure 7.1
'Black box' systems
model

Inputs   Function   Outputs

It is important to try to ensure that all the relevant inputs and outputs are listed. They can all usually be classified as flows of materials, energy or information, and these same classifications can be used to check if any input or output type has been omitted.

Break down the overall function into a set of essential subfunctions

Usually, the conversion of the set of inputs into the set of outputs is a complex task inside the black box, which has to be broken down into subtasks or subfunctions. There is no really objective, systematic way of doing this; the analysis into subfunctions may depend on factors such as the kinds of components available for specific tasks, the necessary or preferred allocations of functions to machines or to human operators, the designer's experience, and so on.

In specifying subfunctions it is helpful to ensure that they are all expressed in the same way. Each one should be a statement of a verb plus a noun: for example, 'amplify signal', 'count items', 'separate waste', 'reduce volume'.

Each subfunction has its own input(s) and output(s), and compatibility between these should be checked. There may be 'auxiliary subfunctions' that have to be added but which do not contribute directly to the overall function, such as 'remove waste'.

Draw a block diagram showing the interactions between subfunctions

A block diagram consists of all the subfunctions separately identified by enclosing them in boxes and linked together by their inputs and outputs so as to satisfy the overall function of the product or device that is being designed. In other words, the original black box of the overall function is redrawn as a 'transparent box' in which the necessary subfunctions and their links can be seen (Figure 7.2).

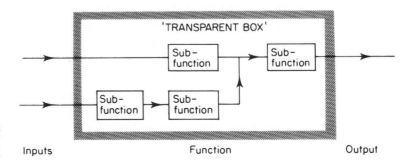

Figure 7.2
'Transparent box'
model

In drawing this diagram you are deciding how the internal inputs and outputs of the subfunctions are linked together so as to make a feasible, working system. You may find that you have to juggle inputs and outputs, and perhaps redefine some subfunctions so that everything is connected together. It is useful to use different conventions, i.e. different types of lines, to show the different types of inputs and outputs, i.e. flows of materials, energy or information.

Draw the system boundary

In drawing the block diagram you will also need to make decisions about the precise extent and location of the system boundary. For example, there can be no 'loose' inputs or outputs in the diagram except those that come from or go outside the system boundary.

It may be that the boundary now has to be narrowed again, after its earlier broadening during consideration of inputs, outputs and overall function. The boundary has to be drawn around a subset of the functions that have been identified, in order to define a feasible product. It is also probable that this drawing of the system boundary is not something in which the designer has complete freedom – as likely as not, it will be a matter of management policy or client requirements. Usually, many different system boundaries can be drawn, defining different products or solution types.

Search for appropriate components for performing the subfunctions and their interactions

If the subfunctions have been defined adequately and at an appropriate level, then it should be possible to identify a suitable component for each subfunction. This identification of components will depend on the nature of the product or device, or more general system, that is being designed. For instance, a 'component' might be defined as a person who performs a certain task, a mechanical component or an electronic device. One of the interesting design possibilities opened up by electronic devices such as microprocessors is that these can often now be substituted for components that were previously mechanical devices or perhaps could only be done by human operators. The function analysis method is a useful aid in these circumstances because it focuses on functions, and leaves the physical means of achieving those functions to this later stage of the design process.

Summary

The aim of the function analysis method is to establish the functions required, and the system boundary, of a new design. The procedure is as follows:

1. Express the overall function for the design in terms of the conversion of inputs into outputs. The overall, 'black box' function should be broad – widening the system boundary.

2. Break down the overall function into a set of essential subfunctions. The subfunctions comprise all the tasks that have to be performed inside the black box.

3. Draw a block diagram showing the interactions between subfunctions. The black box is made 'transparent', so that the subfunctions and their interconnections are clarified.

4. Draw the system boundary. The system boundary defines the functional limits for the product or device to be designed.

5. Search for appropriate components for performing the subfunctions and their interactions. Many alternative components may be capable of performing the identified functions.

Examples

Example 1: feed delivery system

The function analysis method is particularly relevant in the design of flow-process systems, such as that shown diagrammatically in Figure 7.3. This represents a factory where animal feedstuffs are bagged.

In this example, the company wanted to try to reduce the relatively high costs of handling and storing the feedstuffs. A designer might tackle this task by searching for very direct ways in which each part of the existing process might be made more cost-effective. However, a broader formulation of the problem – the overall function – was represented in the following stages:

1. Transfer of feed from mixing bin to bags stored in warehouse.

2. Transfer of feed from mixing bin to bags loaded on truck.

3. Transfer of feed from mixing bin to consumers' storage bins.

4. Transfer of feed ingredients from source to consumers' storage bins.

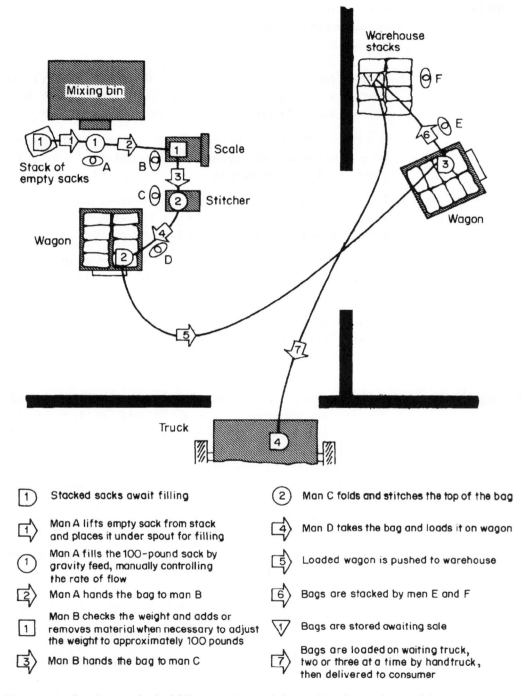

Figure 7.3   Existing method of filling, storing and dispatching bags of animal feed

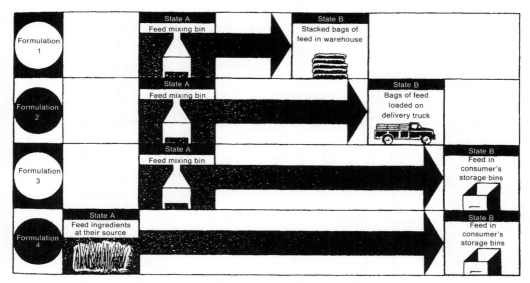

Figure 7.4   Alternative formulations of the feed distribution problem

This broadening of the problem formulation is shown diagrammatically in Figure 7.4.

Each different formulation suggests different kinds of solutions, with the broadest formulation perhaps leading to the complete elimination of the handling, storing and loading subfunctions. (Source: Krick, 1976.)

Example 2: packing carpet squares

This example shows another flow process: the packing of loose carpet squares into lots. The designers first broke down the overall function into a series of principal subfunctions (Figure 7.5).

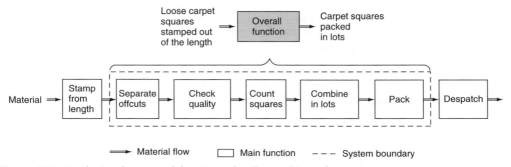

Figure 7.5   Analysis of principal functions for the packing of carpet squares

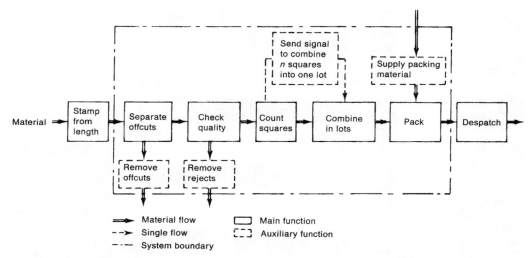

Figure 7.6   Expanded function analysis for the packing of carpet squares

Some auxiliary functions then became clear. For example, the input from the separate stamping machine includes off-cuts which have to be removed; reject squares must also be removed; materials must be brought in for packaging. The subfunction 'count squares' could also be used to give the signal for packaging lots of a specified number (see Figure 7.6). (Source: Pahl and Beitz, 1999.)

Example 3: inkjet printer

Function analysis can be used to identify clusters of subfunctions that can be formulated as distinct modules of a product. This allows modularization of the product design, with the design of separate modules perhaps being delegated to separate designers or specialist teams. It also allows the updating of certain modules during a product's life, without having to change the design of the whole product.

Clusters are groups of subfunctions that depend on each other, can be resolved together and can be reasonably isolated from other clusters. The interactions between clusters therefore have to be minimized, and kept as simple as possible.

Figure 7.7 shows a function analysis for a desktop inkjet printer, with subfunctions clustered. These clusters identify four primary modules for the printer: electronics, ink system (print cartridge), chassis and paper-handling system. These modules depend, for the most part, on just one or two distinct connections between them. (Source: Otto and Wood, 2001.)

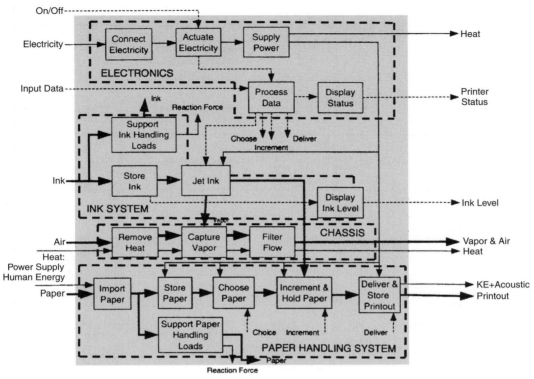

Figure 7.7 Function analysis for an inkjet printer

Example 4: fuel gauge

Function analysis can also be applied even in the design of much smaller products or devices. Figure 7.8 shows the step-by-step development of a function analysis for a fuel gauge. Notice how auxiliary functions are introduced in order to cope with a gradually broadening problem formulation to provide for fuel containers of different sizes and shapes, etc. Figure 7.8 also shows how the system boundary can be drawn in different places, in this case depending on whether the output signal is to be to already existing instruments or whether such an instrument is to be included as an integral part of the design. (Source: Pahl and Beitz, 1999.)

Worked example: washing machine

A relatively simple example of the use of the function analysis method is based on the domestic washing machine. The overall function of such a machine is to convert an input of soiled clothes into an output of clean clothes, as shown in Figure 7.9.

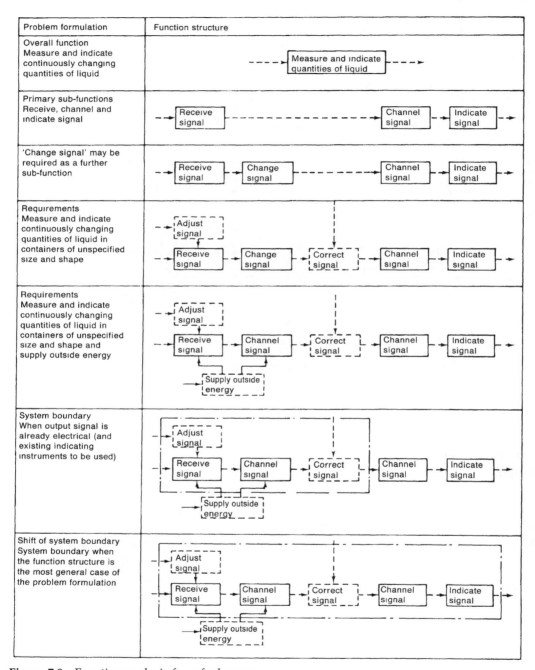

Figure 7.8 Function analysis for a fuel gauge

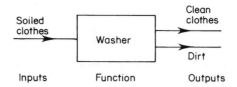

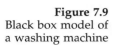

Figure 7.9
Black box model of
a washing machine

Inputs   Function   Outputs

Inside the 'black box' there must be a process that separates the dirt from the clothes, and so the dirt itself must also be a separate output. We know that the conventional process involves water as a means of achieving this separation, and that a further stage must therefore be the conversion of clean (wet) clothes to clean (dry) clothes. Even further stages involve pressing and sorting clothes. The inputs and outputs might therefore be listed like this:

Inputs	*Outputs*	
Soiled clothes	(Stage 1)	Clean clothes
		Dirt
	(Stage 2)	Dry clothes
		Water
	(Stage 3)	Pressed clothes

The essential subfunctions, together with the conventional means of achieving them, for converting soiled clothes into clean, pressed clothes would therefore be as follows:

Essential subfunctions	*Means of achieving subfunctions*
Loosen dirt	Add water and detergent
Separate dirt from clothes	Agitate
Remove dirt	Rinse
Remove water	Spin
Dry clothes	Blow with hot air
Smooth clothes	Press

A block diagram, with main and subsidiary inputs and outputs, might look like that in Figure 7.10. The development of washing machines has involved progressively widening the system boundary, as shown in the figure. Early washing machines simply separated the dirt from the clothes, but did nothing about removing the excess water from the clothes; this was left as a task for the human operator, using either hand or mechanical wringing of the clothes. The inclusion of a spin-drying function removed the excess water, but still left a drying process. This is now incorporated in

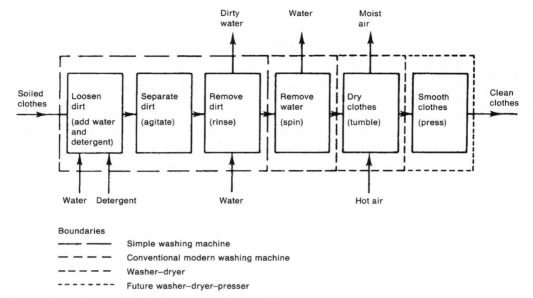

Figure 7.10   Function analysis of a washing machine

washer-driers. Perhaps the smoothing of clothes will somehow be incorporated in future machines. For example, some smoothing function is beginning to be incorporated into tumble-dryers, including the application of steam to remove wrinkles.

8 Setting Requirements

Design problems are always set within certain limits or constraints. One of the most important limits, for example, is that of cost: what the client is prepared to spend on a new machine, or what customers may be expected to pay as the purchase price of a product. Other common limits may be the acceptable size or weight of a machine; some limits will be performance requirements, such as an engine's power rating; still others might be set by statutory legal or safety requirements.

This set of requirements comprises the performance specification of the product or machine. Statements of design objectives or functions (such as those derived from objectives tree or function analysis methods) are sometimes regarded as being performance specifications, but this is not really correct. Objectives and functions are statements of what a design must achieve or do, but they are not normally set in terms of precise limits, which is what a performance specification does.

In setting limits to what has to be achieved by a design, a performance specification thereby limits the range of acceptable solutions. Because it therefore sets the designer's target range, it should not be defined too narrowly. If it is, then a lot of otherwise acceptable solutions might be eliminated unnecessarily. On the other hand, a specification that is too broad or vague can leave the designer with little idea of the appropriate direction in which to aim. Specification limits that are set too wide can also lead to inappropriate solutions which then have to be changed or modified when it is found that they actually fall outside of acceptable limits.

Engineering Design Methods: Strategies for Product Design   N. Cross
© 2008 John Wiley & Sons, Ltd

So there are good reasons for putting some effort into an accurate performance specification early in the design process. Initially, it sets up some boundaries to the 'solution space' within which the designer must search. Later on in the design process, the performance specification can be used in evaluating proposed solutions, to check that they do fall within the acceptable boundaries.

The *performance specification* method is intended to help in defining the design problem, leaving the appropriate amount of freedom so that the designer has room to manoeuvre over the ways and means of achieving a satisfactory design solution. A specification defines the required *performance*, and not the required *product*. The method therefore emphasizes the performance that a design solution has to achieve, and not any particular physical components which may be means of achieving that performance.

The Performance Specification Method

Procedure

Consider the different levels of generality of solution which might be applicable

It is important that a specification is addressed to an appropriate level of generality for the solution type that is to be considered. A specification at too high a level of generality may allow inappropriate solutions to be suggested, whereas too low a level (a specification which is too specific) can remove almost all of the designer's freedom to generate a range of acceptable solutions.

So the first step is to consider the different levels of generality. A simple classification of types of level, from the most general down to the least, for a product might be:

- product *alternatives*

- product *types*

- product *features*.

As an example to illustrate these levels, suppose that the product in question is a domestic heating appliance. At the highest level of generality the designer would be free to propose alternative ways of heating a house, such as moveable appliances, fixed appliances, central heating with radiators, ducted warm air, etc. There might even be freedom to move away from the concept of an 'appliance' to alternative forms of heating such as conservatories that trap solar heat; or to ways of retaining heat, such as insulation. At the intermediate level, the designer would have a much more limited freedom, and might only be concerned with different types of appliance, say, different heater types such as radiators or convectors, or different fuel types. At the lowest level, the designer would be constrained to considering different features within a particular type of appliance, such as its heating element, switches, body casing, supports, etc.

Determine the level of generality at which to operate

Considering the different levels of generality might lead either to a broadening or a narrowing of initial product concepts or of the design brief. The second step of the method is therefore to make a decision on the appropriate level.

Normally, the client, company management or customer decides the level at which the designer will operate. For instance, in the case of domestic heating appliances, the highest level of generality ('alternatives') would only be considered if an appliance manufacturer was proposing to diversify or broaden its activities into other aspects of domestic heating. The intermediate level ('types') would normally be considered when a new product was to be designed, to add to the existing range of appliances or to replace obsolete ones. The lowest level ('features') would be considered when making modifications to existing products.

The higher the level of generality that may be considered, then the more freedom the designer has in terms of the range of acceptable solutions. Of course, the higher levels also subsume the lower levels of specification; that is, the specification of features is part of the specification of types which is part of the specification of alternatives.

Identify the required performance attributes

Once the level at which designing is to proceed has been decided, work can begin on the performance specification proper. Any product or machine will have a set of *attributes*, and it is these which are specified in the performance specification. Attributes include such things as comfort, portability and durability, and key features such as speed, cost and safety.

Performance attributes are usually similar to, or derived from, the design objectives and functions. So if you have already prepared an objectives tree or a functions analysis, these are likely to be the source of your initial list of performance attributes.

A most important aspect to bear in mind when listing performance attributes is that they should be stated in a way which is independent of any particular solution. Statements of attributes made by clients or customers are often couched in terms of solutions, because they value some performance aspect which is embodied in the solution but they have not separated the attribute from a particular embodiment. Such solution-based rather than performance-based statements are usually unnecessarily restrictive of solution concepts.

For example, a client might suggest that the material for a particular surface area should be ceramic tiles, because that is a satisfactory feature of an existing solution. But the essential performance requirement might be that the surface should be nonporous, or easy to clean, or have a smooth, hard texture, or simply have a shiny appearance. Acceptable alternatives might be plastics, metal or marble.

There may be a whole complex of reasons underlying a client or customer specification of a particular solution feature. It could be the whole set of attributes of a ceramic surface, as just listed, plus the mass which is provided by ceramic tiles, plus the colour range, plus some perceived status or other value which is not immediately obvious. A comprehensive and reliable list of performance attributes can therefore take some considerable effort to compile, and may well require careful research into client, customer and perhaps manufacturer requirements.

The final list of performance attributes contains all the conditions that a design proposal should satisfy. However, it may become necessary to distinguish within this list between those attributes or requirements that are 'demands' and those that are 'wishes'. Demands are requirements that must be met, whereas wishes are those that the client, customer or designer would like to meet if possible. For example, the requirement of a nonporous surface might be a functional 'demand', but availability in a range of colours might be a 'wish' dependent on the material actually chosen.

State succinct and precise performance requirements for each attribute

Once a reliable list of attributes has been compiled, a performance specification is written for each one. A specification says what a product must *do*; not what it must *be*. Again, this may well require some careful research – it is not adequate simply to guess at performance requirements, nor just to take them from an existing solution type.

Wherever possible, a performance requirement should be expressed in quantified terms. Thus, for example, a maximum weight should be specified, rather than a vague statement such as 'lightweight'. A safety requirement – say, for escape from a vehicle – should state the maximum time allowable for escape in an emergency, rather than using terms like 'rapidly' or 'readily'.

Also, wherever possible and appropriate, a requirement should set a range of limits within which acceptable performance lies. So a requirement should not say 'seat height: 425 mm' if a range between 400 mm and 450 mm is acceptable. On the other hand, spurious 'precision' is also to be avoided: do not specify 'a container of volume 21.2 l' if you mean to refer to a wastepaper bin of 'approximately 300 mm diameter and 300 mm high'.

Summary

The aim of the performance specification method is to make an accurate specification of the performance required of a design solution. The procedure is as follows:

1. Consider the different levels of generality of solution which might be applicable. There might be a choice between

   - product alternatives

   - product types

   - product features.

2. Determine the level of generality at which to operate. This decision is usually made by the client. The higher the level of generality, the more freedom the designer has.

3. Identify the required performance attributes. Attributes should be stated in terms which are independent of any particular solution.

4. State succinct and precise performance requirements for each attribute. Wherever possible, specifications should be in quantified terms, and identify ranges between limits.

Examples

Example 1: one-handed mixing tap

This example is a specification for a domestic water mixing tap that can be operated with one hand (Figure 8.1). The initial design brief for this project is given in Chapter 1. Notice how the brief has been considerably expanded, as the design team has researched the problem. Some details in the brief have changed (for example, the maximum pressure) as a result of establishing the national standards that apply to such a product. The range of users has also been taken into account (requirement 11: light operation for children), as have safety considerations (requirements 18–20). The project timescale has also been included in the specification. The 'D or W' column

		Specification		
		for	One-handed mixing tap	*Page* 1

Changes	D or W	Requirements	Responsible
	D	1 Throughput (mixed flow) max. 10 l/min at 2 bar	
	D	2 Max. pressure 10 bar (test pressure 15 bar as per DIN 2401)	
	D	3 Temp. of water: standard 60°C, 100°C (short-time)	
	D	4 Temperature setting independent of throughput and pressure	
	W	5 Permissible temp. fluctuation ±5°C at a pressure diff. of ±5 bar between hot and cold supply	
	D	6 Connection: 2 × Cu pipes, 10 × 1 mm, l = 400 mm	
	D	7 Single-hole attachment $\diameter$ 35$^{+2}_{-1}$ mm, basin thickness 0 – 18 mm (Observe basin dimensions DIN EN 31, DIN EN 32, DIN 1368)	
	D	8 Outflow above upper edge of basin: 50 mm	
	D	9 To fit household basin	
	W	10 Convertible into wall fitting	
	D	11 Light operation (children)	
	D	12 No external energy	
	D	13 Hard water supply (drinking water)	
	D	14 Clear identification of temperature setting	
	D	15 Trade mark prominently displayed	
	D	16 No connection of the two supplies when valve shut	
	W	17 No connection when water drawn off	
	D	18 Handle not to heat above 35°C	
	W	19 No burns from touching the fittings	
	W	20 Provide scalding protection if extra costs small	
	D	21 Obvious operation, simple and convenient handling	
	D	22 Smooth, easily cleaned contours, no sharp edges	
	D	23 Noiseless operation (≤20 dB as per DIN 52218)	
	W	24 Service life 10 years at about 300 000 operations	
	D	25 Easy maintenance and simple repairs. Use standard spare parts	
	D	26 Max. manuf. costs DM 30 (3000 units per month)	
	D	27 Schedules from inception of development	

	Conceptual design	Embodiment design	Detail Design	Prototype	
after	2	4	6˙	9	months

Figure 8.1 Specification for a one-handed mixing tap

on the left distinguishes between 'Demands' and 'Wishes' in the specification. (Source: Pahl and Beitz, 1999.)

Example 2: fuel gauge

This problem was formulated by the client at the lowest level of generality: the design of a particular type of fuel gauge for use in motor vehicles. The initial general formulation of the problem statement was:

> A gauge to measure continuously changing quantities of liquid in containers of unspecified size and shape, and to indicate the measurement at various distances from the containers.

The following list of attributes was then developed:

- Suitable for containers (fuel tanks) of

  - various volumes

  - various shapes

  - various heights

  - various materials.

- Connection to top or side of container.

- Operates at various distances from container.

- Measures petrol or diesel liquid.

- Accurate signal.

- Reliable operation.

The design team went on to develop a full performance specification, as shown in Figure 8.2. As in the previous example, they also distinguished between 'Demands' (D) and 'Wishes' (W). (Source: Pahl and Beitz, 1999.)

Example 3: electric toothbrush

This example shows the development of a performance specification for a consumer product: an electric toothbrush. The problem is set at the intermediate level of generality, i.e. a new type of toothbrush, but it has novel features which require precise performance specifications.

		Specification		
		for Fuel gauge		*Page 1*

Changes	*D or W*	*Requirements*	*Responsible*
		1. Container, connection, distance	
	D	Volume: 20 – 160 l	
		Shape fixed or unspecified (rigid)	
	D	Material: steel or plastic	
		Connection to container:	
	W	Flange connection	
	D	Top connection	
	D	Side connection	
		H = 150 – 600 mm	
	W	d = o 71 mm. h = 20 mm	
	D	Distance from container to indicator:	
		≠ 0 m, 3 – 4 m	
	W	1 – 20 m	
		2. Contents, temperature range, material	
		Liquid Operating range Storage environment	
	D	Petrol or diesel – 25 to +65°C – 40 to + 100°C	
		3. Signal, energy	
	W	Output of transmitter: electric signal (voltage change with quantity change)	
	D	Available source of energy: d.c. at 6, 12, 24 V	
		Voltage variation — 15 to +25%	
	D	Output signal accuracy at max. ±3%	
	W	±2%	
		(together with indicator error ±5%)	
		under normal conditions, horizontal level. v = constant;	
		able to withstand shocks of normal driving	
	D	Response sensitivity: 1% of maximum output signal	
	W	0.5% of maximum output signal	
	D	Signal unaffected by angle of liquid surface	
	D	Possibility of signal calibration	

Figure 8.2   Specification for a fuel gauge

		Specification	
		for Fuel gauge	Page 2

Changes	D or W	Requirements	Responsible
	W	Possibility of signal calibration with full container	
	D	Minimum measurable content: 3% of maximum value	
	W	Reserve tank contents by special signal	
		4. Operating conditions	
	D	Forward acceleration ± 10 m/s²	
	D	Sideways acceleration ± 10 m/s²	
	D	Upward acceleration (vibration) up to 30 m/s²	
	W	Shocks in forward direction without damage up to 30 m/s²	
	D	Forward tilt up to ± 30°	
	D	Sideways tilt max. 45°	
	D	Tank not pressurized (ventilated)	
		5. Test requirements	
	D	Salt spray tests for inside and outside components according to client's requirements	
	D	Pressure test for container 30 kN/m²	
		6. Life expectancy, durability of container	
	D	Life expectancy 5 years in respect of corrosion due to contents and condensation	
	D	Must conform with heavy vehicle requirements	
		7. Production	
	W	Simply modified to suit different container sizes	
		8. Operation, maintenance	
	W	Installation by non-specialist	
	D	Must be replaceable and maintenance-free	
		9. Quantity	
		10 000/day of the adjustable type, 5000/day of the most popular type	
		10. Costs	
		Manufacturing costs ⩽ DM 3.00 each	

Figure 8.2   (Continued)

The designers listed the new product's attributes mainly in terms of a set of user needs at the concept stage:

Physiological needs	Clean teeth better than a conventional toothbrush, massage gums, reduce decay, hygienic family sharing, electrical and mechanical safety, etc.
Social needs	Sweet breath and white teeth (symbolic needs for social acceptance); handle colours to match bathroom, etc.
Psychological needs	Autonomy in deciding when and how one's teeth are to be cared for, self-esteem from care of teeth, praise for effort, pleasure from giving or receiving a gift, etc.
Technical needs	Diameter, length, brush size, amplitude, frequency, weight, running time, reliability, useful life, etc.
Time needs	Needed for Christmas market.
Resources exchanged	$1 per person is the lowest cost alternative, but electric razors sell for 20 times the price of a manual razor, so probably $20 will be paid for an electric toothbrush.

The performance specification was then drawn up as a set of design objectives with corresponding criteria, as shown in Figure 8.3. (Source: Love, 1986.)

Example 4: seat suspension unit

The design task in this example was to modify the design of seats in industrial vehicles, such as excavators and loaders, so as to incorporate a suspension mechanism. Historically, machines of this type did not have much suspension other than that provided by the vehicle's pneumatic tyres. However, drivers of such vehicles – construction vehicles, farm tractors, etc. – have been frequently found to suffer back troubles and injuries. In some cases, injuries were worsened by the need for the driver to turn in the seat so as to observe lateral and rearward as well as forward operations. The design objectives were therefore to provide a suspension mechanism for the operator's seat, to dampen vibrations to within acceptable limits and to allow the position and orientation of the seat to be fully adjustable.

Figure 8.4 shows the performance specification developed for this problem. Note that there are some general industry standards to be observed, as well as particular requirements for the use of this type

Objectives	Criteria
1. To be attractive, suitable for sale primarily in the gift market and secondly as a personal purchase.	1a. Attractiveness of overall design and packaging to be judged better than brands X and Y by more than 75% of a representative consumer panel.
	1b. Decorator colours to be the same as our regular products.
	1c. Package can be displayed on counter area of 75 × 100 mm.
2. The technical functions are to be at least as good as past 'family' models of brand X.	2a. Technical functions to be judged at least as good as the past 'family' model of brand X by dental consultant, Dr J.P.
	2b. Amplitude to be between 2 and 3 mm.
	2c. Frequency to be 15 ± 5 cycles/s.
	2d. Battery life to be minimum of 50 min. when tested according to standard XYZ.
	2e, etc., for other technical aspects such as weight, impact strength, frequency of repair, dimensions . . . .
3. To be saleable in the United States and Canada.	3. Must meet UL and CSA standards for safety (a crucial criterion).
4. The timing objective is that the product be ready for sale to the Christmas trade in the nearest feasible season.	4. The time milestones, backing up from October production are to be: • mock-up approval—2 months • tooling release—6 months • production prototype—10 months • pilot run—10 months • production run—13 months (October).
5. The selling price is to be not more than 10% of the present utility models.	5. The selling price is to be between $12.50 and $17.50, depending on the features offered, for a production run of 100 000 units.

Figure 8.3 Performance specification for an electric toothbrush

of seat. Environmental considerations are taken into account, both in terms of the environment in which the mechanism functions and in terms of disposal at the end of its useful life. (Source: Hurst, 1999.)

Worked example: portable fax machine

Communication devices of various kinds have proliferated, especially with the growing use of radio and satellite communications links, data transfer by telephone, Internet communications, etc. Many communications devices first appeared as large, office-based,

DEFINITIONS:

PERFORMANCE REQUIREMENTS:

The mechanism must allow full adjustment of the seat position. To comply with ISO 4253 these adjustments are rotate through 180 degrees in either direction, 80 mm up and down in the vertical plane and 150 mm front and back in the horizontal plane.

Increments of adjustment must be less than 30 degrees and 25 mm respectively.

The natural frequency of vibration of the combined seat and operator must be <2.5 Hz. Isolation criteria for class 3 seats as set by ISO 7096.

The mechanism must still operate with the machine on a 30 degree slope in any direction.

Suspension travel must be vertical and a maximum of 110 mm. Amplitudes must be limited under resonant conditions and step inputs.

The temperature range during operation is between −10 and +50°C and whilst stored could drop to −30°C.

The humidity will range from 0 to 80%.

The suspension mechanism will also be subjected to rain, snow and heavy organic and mineral grime.

The ex-works cost of the mechanism must be <£30.

The target population of operators is to be restricted to people between the ages of 19 and 65. Sizes, weights and strengths are to be between the 5th and 95th percentiles. For example, adjustment must accommodate drivers in the weight range of 60 to 130 kg.

The quality of the mechanism must be consistent with the rest of the machine.

The required design life is 10 000 hours of operation.

The required reliability is 90% over the 10 000 hours of operation.

The appearance must be as rugged as the rest of the machine.

The weight of the complete mechanism must be <50 kg.

The maximum overall size is 0.5 × 0.5 × 0.5m. In the horizontal plane the mechanism must have a radius about the centre of rotation <300 mm.

The mechanism must be capable of being fitted to the full range of seat bases and machine floors.

MANUFACTURE REQUIREMENTS:

The machine will be assembled on a ten stage assembly line. The mechanism will be assembled prior to installation as far as is possible. Installation must take <20 minutes.

The mechanism is to be manufactured and finished in-house.

Any materials can be used as long as they comply with other statements in this specification.

6000 are to be produced each year.

ACCEPTANCE STANDARDS:

In accordance with ISO 3776 anchorage points must be provided for seat belts which accept a pull load through the suspension of 5000 lbs.

Every mechanism will be inspected prior to assembly in the machine.

Accelerated cyclic tests of five fully loaded mechanisms are to be carried out to verify the reliability levels and fatigue strength.

The mechanism must not conflict with existing patents.

DISPOSAL:

The suspension mechanism must not contain any hazardous materials and all polymeric materials used must be clearly identified.

OPERATION REQUIREMENTS:

Adjustment of the seat position or the level of damping must be easily carried out by the operator whilst in the sitting position in <30 seconds.

Removal of the mechanism from the machine by one person must be possible in <30 minutes.

The device is to be maintenance free for the life of the machine.

Secure locking must be provided after adjustment.

Seat movement and locking must be fail safe.

It must not be possible for the operator to trap their fingers in the mechanism.

Figure 8.4 Specification for a seat suspension mechanism

immobile machines, and then gradually became smaller, lighter and portable. Telephones and computers are classic examples; so too is the fax machine. Some people now need to have not only one at the office and one at home, but also one that can be used at other locations and that therefore travels with them. This example is based on the design of such a portable fax machine.

The initial level of generality for this problem has been set by the client's request for the design of a new, portable fax machine: it is a particular product *type*, and so the designer has some freedom to generate new product ideas, rather than just being constrained to product *features*.

There are many specialized attributes which would have to be researched and specified, such as the industry communication standards, scanning and printing devices to be incorporated, etc. We shall concentrate here primarily on the key attribute of 'portable'. What exactly does this mean? We need to know what features of 'portability' might be important to potential purchasers and users of the fax machine.

We therefore interview a range of fax machine users, and potential new users of a portable facility, about their needs. Typical users for a portable fax machine are business representatives, engineers or others who have to travel in their work and communicate with their head office or other locations by means of documents such as drawings, order forms, etc. From this it emerges that there are two distinct aspects to portability. The first is, quite simply, that the machine can be carried and used comfortably and easily. The second aspect is that the purpose of a portable machine is that it can be used in a wide variety of different locations (e.g. clients' offices, construction sites and suppliers' factories). The portability attribute is therefore strongly related to usability of the machine in such environments.

Further research with users is necessary to develop performance specifications for both of these aspects of portability. For example, to specify the 'carryable' performance features, it is not adequate simply to suggest a carrying handle. Nor is it adequate just to weigh a rival product and specify that as a maximum weight. We need to know the range of users for the fax machine and the typical distances or lengths of time that it might be carried. Experiments with a few representative, least-strong users and maximum expected carrying times could then establish an appropriate weight limit.

We also need to investigate further the variety of locations in which it is desired to use the fax machine. One aspect of the typical use of such a machine is that it is not always possible to use it on a desk or other stable surface. Sometimes, as with portable computers, use includes on someone's lap on a train or in an airport lounge. Therefore the machine must be small but stable. Does some potential use include during meetings or conferences (e.g. by journalists faxing press releases)? In that case its operation should be silent or very quiet. Does it include out-of-doors use? In that case there might be weatherproofing requirements, or the user might be wearing gloves, with implications for the design of buttons, controls, paper feed mechanisms, etc.

Obviously, in many locations there is no available power source, and so a portable fax machine must have its own batteries. But it may be that salespeople and others often use the fax machine in their car, and therefore the car's battery could be used through the cigarette-lighter socket. Another aspect of performance that emerges is that use of the fax machine will very often be in conjunction with a mobile telephone, and therefore must have connectors to enable them to plug into these as well as into conventional telephone sockets. (And there is no need for the fax machine itself to incorporate a telephone.) One other aspect to emerge from discussion with potential users is that the fax machine could be useful in conjunction with a portable computer, both as a scanner of documents for entering into the computer and as a printer. Appropriate sockets and connectors are therefore also necessary for this.

In fact, it transpires that the concept of a 'portable fax machine' can be rethought as a portable modem/scanner/printer for use in conjunction with a laptop computer and/or mobile telephone. It may be that the designers will want to suggest to the client something that is a kind of new product *alternative*, rather than being just a new product *type*.

An outline performance specification for the 'portability' attribute is therefore developed as follows:

- Can be carried in one hand; preferably has a carrying handle.

- Weight not more than 4 kg, including batteries.

- Optional carrying case, with pockets for power and connector cables.

- Maximum base dimensions: 300 mm × 300 mm.

- Operating environment ranges: temperature, +1 °C to +35 °C; relative humidity, 20 % to 70 %.

- Weatherproof against rain showers when not in use.

- Silent in use: no warning/function bleepers, etc.

- Displays and controls legible in low-light environments.

- Compatible with fixed and mobile telephones.

- Compatible with portable computers.

- Power sources:

  - mains

  - own battery

  - car cigarette-lighter socket.

9 *Determining Characteristics*

In determining a product specification, conflict and misunderstanding can sometimes arise between the marketing and the engineering members of the design team. This is usually because they focus on different interpretations of what should be specified. Managers and market researchers tend to concentrate more on specifying the desirable *attributes* of a new product (usually from the viewpoint of customer or client requirements), whereas designers and engineers concentrate more on a product's engineering *characteristics* (usually in terms of its physical properties).

The relationship between characteristics and attributes is in fact a very close one, and confusion can be avoided if this relationship is clearly understood. Designers make decisions about the product's physical properties, and thus determine its engineering characteristics; but those characteristics then determine the product's attributes, which in turn satisfy customers' needs and requirements. Thus, for example, the engineering designer may choose a particular metal casing for a product, of a certain gauge and surface finish, thus determining characteristics such as weight, rigidity and texture; these characteristics determine product attributes such as portability, durability and appearance.

With increased competition in all product markets, it has become necessary to ensure that this relationship between engineering characteristics and product attributes is properly understood. In particular, it is necessary to understand just what customers want in terms of product attributes and to ensure that these are carefully translated into specifications of the appropriate engineering characteristics. This attitude towards product design is based on the

Engineering Design Methods: Strategies for Product Design   N. Cross
© 2008 John Wiley & Sons, Ltd

philosophy of 'listening to the voice of the customer', and is reflected in an increased concentration on product quality. Design for quality is recognized as a major factor in determining the commercial success of a product.

A comprehensive method for matching customer requirements to engineering characteristics is the *quality function deployment* (QFD) method. 'Quality function deployment' is a direct translation of the Japanese characters *Hin Shitsu, Ki No, Ten Kai*. In Japanese, the phrase means something like the strategic arrangement (deployment) throughout all aspects of a product (functions) of appropriate characteristics (qualities) according to customer demands.

The QFD method recognizes that the person who buys (or who most influences the buying decision for) a product is the most important person in determining the commercial success of a product. If customers do not buy it, then the product – however 'well designed' it may be – will be a commercial failure. Therefore 'the voice of the customer' has priority in determining the product's attributes. This means taking care to identify who the customers are, to listen carefully to what they say and to determine the product's engineering characteristics in the light of this.

As it is presented here, QFD is essentially concerned with the translation of customer requirements into engineering characteristics. However, because it is a comprehensive method, aspects of QFD can be used at various stages of the design process, and it also draws upon features from several other design methods.

The Quality Function Deployment Method

Procedure

Identify customer requirements in terms of product attributes

The method starts with the identification of the customers and of their own views of their requirements and desired product attributes. There are various market research techniques that can be used to assist the gathering of information about customer requirements and preferences. These methods include product 'clinics' where customers are quizzed in depth about what they like and dislike about particular products, and 'hall tests' where various competing products are arranged on display in a room or hall and customers are asked to inspect the products and give their thoughts and reactions. The use of focus groups, questionnaires, etc., was introduced in the user scenarios method, Chapter 5.

Usually, of course, customers will talk about products both in terms of general attributes and specific characteristics: observations ranging from 'It's easy to use' to 'I don't like the colour'. As in the performance specification method, it may be necessary to interpret the more general statements into more precise statements of requirements, but it is important to try to identify and to preserve customers' wishes and preferences, rather than to reinterpret their observations into the designer's perceptions of what the customers 'really mean'. For this reason, words and phrases actually used by customers are often retained in statements of product attributes, even though they may seem to be vague and imprecise.

Determine the relative importance of the attributes

Of course, not all the identified product attributes will be equally important to customers. For example, 'easy to use' may be regarded as much more important than 'easy to maintain'. Also, some requirements (as noted in the performance specification method) may be 'demands' or absolute requirements (e.g. 'safe to use') rather than relative preferences.

The design team will want to know which attributes of their product design are ones that most heavily affect customers' perceptions of the product, and so it is necessary to establish the relative importance of those attributes to the customers themselves. Again market research methods can help to establish these relative preferences, and provide confirmation of whether what customers say they want is actually reflected in what they buy.

Some relatively simple techniques can also be used in order to assess the relative importance of the identified attributes. For example, customers can be asked to rank-order their statements of requirements, or to allocate 'points' (preferably from a fixed maximum points allowance) to the various attributes. (Techniques for doing this are discussed in the weighted objectives method, Chapter 11.)

The outcome of this step in the procedure is the allocation of relative 'weights' to the set of customer-specified product attributes. Normally, a percentage value is set for each attribute – i.e. the weights for the complete set of attributes add up to a total of 100.

Evaluate the attributes of competing products

Customers often make judgements about product attributes in terms of comparisons with other products. For example, a car buyer may say that car A 'feels more responsive than car B'. This use of comparisons is perfectly understandable, given that customers are not usually experts and can only guess at what is possible in product design through observation of what some products actually achieve. Market research information is also often collected by methods of comparison between products.

In a competitive market, therefore, the design team has to try to ensure that its product will satisfy customer requirements better than the competitor products. The performance of the competition is therefore analysed, particularly with regard to those product attributes that are weighted high in relative importance. Some of these performance measures will be objective and quantitative, whilst some will be subjective comparisons as made by customers. However, even when objective measures can be made, these should be checked against customers' perceptions, which may not correspond with the objective measures.

In designing a new product, there may not be many competitor products, but that would be unusual; most product designs have to compete against existing products already on the market. In those cases where a design team is redesigning or improving an existing product, this step in the procedure not only highlights where improvements to the design team's product may be necessary, but also where this current product already has advantages over the competition, which should be maintained. The performance scores for the team's own current product and for the competition should be listed against the set of product attributes.

Draw a matrix of product attributes against engineering characteristics

As suggested above, customers are not experts and therefore cannot usually specify their requirements in terms of the product's engineering characteristics that influence those requirements. For example, the car buyer may know what 'responsiveness' feels like, but is unlikely to be able to refer to this in terms of engine torque. It is therefore necessary for the design team to identify those engineering characteristics of their product that satisfy or influence in any way the customer requirements. For instance, the overall weight of a car, as well as its engine torque, will influence its 'responsiveness'.

The engineering characteristics must be real, measurable characteristics over which the engineering designer has some control. It is understandable for customers to be rather vague about their requirements, or to express them in frustratingly subjective terms, but the engineering designer can only work with the quantitative parameters of identifiable engineering characteristics. It is through the adjustment of the parameters of those characteristics that the designer influences the performance and/or customers' perception of the product. Therefore it is often necessary to put considerable effort into identifying the relevant engineering characteristics and ensuring that each of these can be expressed in measurable units.

Of course, not all engineering characteristics affect all product attributes, and drawing up a matrix will enable the team to identify which characteristics do affect which attributes. It is usual to list the attributes, together with their relative weights, vertically, down the left edge of the matrix, and the characteristics horizontally, along the top edge. The attributes thus form the rows of the matrix, and the characteristics form the columns. Each cell of the matrix represents a potential interaction or relationship between an engineering characteristic and a customer requirement.

Down the right edge of the matrix can be listed the results of the evaluation of competing products, showing the scores achieved against the product attributes for the competing products and the design team's own current product. Along the bottom edge of the matrix is the usual place for recording the units of measurement of the engineering characteristics. If a product already exists and is being redesigned then the product's own values for these characteristics can also be inserted here, together with values achieved by competitor products.

Identify the relationships between engineering characteristics and product attributes

By checking through the cells of the matrix it is possible to identify where any engineering characteristic influences any product attribute. These relationships between characteristics and attributes will not all be of equal value. That is to say, some characteristics will have a strong influence on some attributes, whilst other characteristics might only have a weak influence.

The design team therefore works methodically through the matrix, and records in the matrix cells wherever a relationship occurs, and the strength of that relationship. Sometimes numbers are used to represent the strength of the relationship (e.g. 6 for a strong relationship, 3 for a medium-strength relationship, 1 for a weak relationship), or symbols can be used. When numbers are used, it is possible to enter a second value in each cell, which is the relative weight of the attribute multiplied by the strength of the relationship. The large scores amongst these values enable the design team easily to identify where the adjustment of engineering characteristics will have a large influence on customers' overall perception of the product. However, unless accurate measures of the strength of the relationships can be established, it must be remembered that there is a spurious accuracy implied by the numbers.

Identify any relevant interactions between engineering characteristics

It is often the case that engineering characteristics interact with each other, particularly in terms of their influence on customers' perceptions of the product. For example, a more powerful engine is also likely to be heavier, thus increasing the vehicle weight, and so not necessarily increasing its perceived 'responsiveness'. These interactions can be either negative or positive.

A simple way of checking these interactions is to add another section to the interaction matrix. This new section is usually added on top of the existing matrix, and because it provides a triangular-shaped 'roof' to the matrix, and thus an overall 'house' appearance, the resulting diagram is often referred to as the 'house of quality'.

Working through the 'roof' matrix enables a systematic check to be made of the interactions between the engineering characteristics, and whether these interactions are negative or positive. However, many assumptions may have to be made about the final design when completing the 'roof' matrix, and it should be remembered that changes in the design concept may result in changes in these interactions.

Set target figures to be achieved for the engineering characteristics

In following this method so far, the design team will already have gained substantial insight into their product design, including customer perceptions of their product and of competing products, and how the engineering characteristics of the product relate to customer requirements. In this step of the procedure, the team determines the targets that can be set for the measurable parameters of the engineering characteristics in order to satisfy customer requirements or to improve the product over its competitors.

Of course, in a competitive situation it is important to know what the competitors achieve on the characteristics of their product, so detailed investigation of competitor products may be necessary. The design team can then set targets for themselves which would be better than the competition. Sometimes it may be necessary to conduct trials with customers in order to determine what would be acceptable target figures to set. This is similar to determining values in a performance specification.

Summary

The aim of the quality function deployment method is to set targets to be achieved for the engineering characteristics of a product, such that they satisfy customer requirements. The procedure is as follows:

1. Identify customer requirements in terms of product attributes. It is important that 'the voice of the customer' is recognized, and that customer requirements are not subject to 'reinterpretation' by the design team.

2. Determine the relative importance of the attributes. Techniques of rank-ordering or points-allocation can be used to help determine the relative weights that should be attached to the various attributes. Percentage weights are normally used.

3. Evaluate the attributes of competing products. Performance scores for competing products and the design team's own product (if a version of it already exists) should be listed against the set of customer requirements.

4. Draw a matrix of product attributes against engineering characteristics. Include all the engineering characteristics that influence any of the product attributes and ensure that they are expressed in measurable units.

5. Identify the relationships between engineering characteristics and product attributes. The strength of the relationships can be indicated either by symbols or numbers; using numbers has some advantages, but can introduce a spurious accuracy.

6. Identify any relevant interactions between engineering characteristics. The 'roof' matrix of the 'house of quality' provides this check, but may be dependent upon changes in the design concept.

7. Set target figures to be achieved for the engineering characteristics. Use information from competitor products or from trials with customers.

Examples

Example 1: bicycle splashguard

This is a relatively simple product, but it illustrates how considerable effort can be necessary in designing to satisfy customer requirements even for simple products. It is a design for a new product type: a detachable splashguard for the rear wheel of mountain bikes. Normally such bicycles have no mudguards, but for circumstances in which the rider does not wish to dirty his or her clothes (from water and mud thrown up at the rear) a detachable splashguard was thought to be a potentially desirable product.

Figure 9.1 shows the QFD interaction matrix prepared for this design problem. The design team interviewed mountain bicycle riders and asked them what features they would like to see incorporated in a detachable splashguard, and then organized this information (retaining 'the voice of the customer' as far as possible) into product attributes. Additional requirements from the sponsor of the project were also added, such as cost, time and manufacturing requirements. These are all listed, in groups, down the left edge of the matrix.

Some of the requirements were 'musts' – i.e. they had to be satisfied absolutely – whereas others could be weighted relative to each other. These relative weights are listed alongside the requirements, and the absolute requirements are marked with an asterisk.

Since there were no comparable products already on the market, the team could not evaluate the attributes of competing products. However, they decided to make comparisons with two other possible means of avoiding dirt being splashed onto the rider: a fixed mudguard and a raincoat. The evaluations of these alternatives are given down the right edge of the matrix.

Engineering characteristics for a splashguard design were then established, related to the desired product attributes. For example, the 'easy to attach' attribute could be measured by (1) the number of steps needed to attach, (2) the time needed to attach, (3) the number of parts needed and (4) the number of standard tools needed. These characteristics (called 'engineer requirements' in the example) are listed along the top of the matrix, and values for the strength of the relationship between characteristics and attributes are shown in the appropriate cells of the matrix.

Finally, along the bottom of the matrix are listed the units by which the engineering characteristics can be measured, and the targets set for the new product design in comparison with those achieved by the alternatives. The design team thus established a

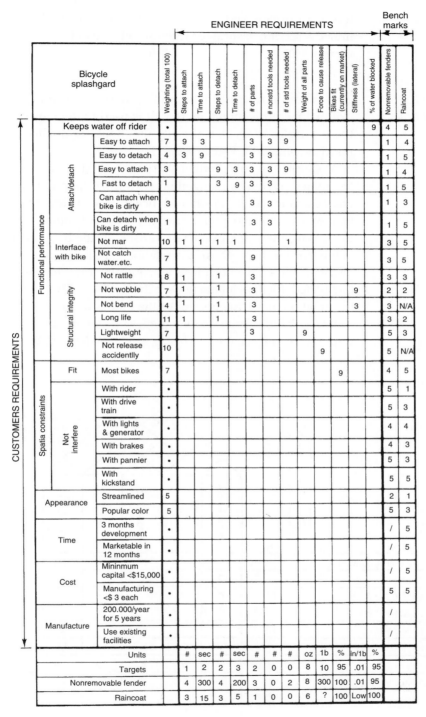

Figure 9.1 Interaction matrix of requirements for a bicycle splashguard

thorough understanding of this novel design problem and determined a set of measurable targets to achieve in their design. (Source: Ullman, 2002.)

Example 2: cordless drill

This example refers to a project to design a hand-held cordless drill for the professional market. It is used to demonstrate some of the principles of QFD, especially the 'house of quality' interaction matrices, shown in Figure 9.2.

A simple list of product attributes is shown (here CAs, or 'customer attributes'), together with corresponding engineering characteristics (ECs). Positive or negative signs are given with the ECs to indicate the preferred desirability of either increasing or decreasing the value of a characteristic. In the body of the matrix, values for the strengths of the relationships between attributes and characteristics have been inserted, and the 'roof' has been partially completed for the interactions between ECs. The supplementary information along the bottom of the matrix includes the EC values achieved by the principal competitor product, and the 'imputed importance' of each EC. This is a value derived by computing

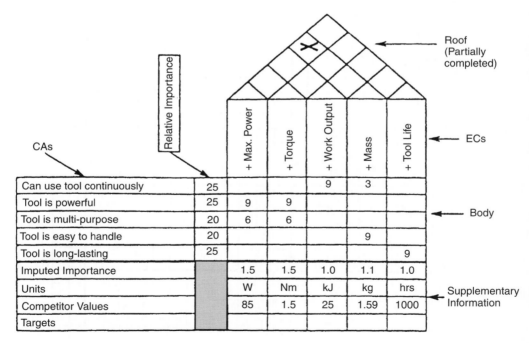

CAs	Relative Importance	+ Max. Power	+ Torque	+ Work Output	+ Mass	+ Tool Life
Can use tool continuously	25			9	3	
Tool is powerful	25	9	9			
Tool is multi-purpose	20	6	6			
Tool is easy to handle	20				9	
Tool is long-lasting	25					9
Imputed Importance		1.5	1.5	1.0	1.1	1.0
Units		W	Nm	kJ	kg	hrs
Competitor Values		85	1.5	25	1.59	1000
Targets						

Figure 9.2   'House of quality' for the design of a cordless drill

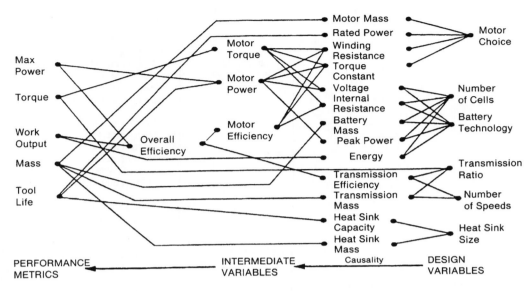

Figure 9.3 Network of physical properties influencing the engineering characteristics of a cordless drill

the weighted sums of EC–CA relationships. These relative values indicate where changing an engineering characteristic may have most influence.

However, the complexities of the engineering characteristics themselves are illustrated in Figure 9.3. This traces the network of properties which influence the engineering characteristics. For example, the 'torque' characteristic is determined by the transmission ratio and the motor torque, which in turn is determined by winding resistance, torque constant, voltage and internal resistance. Thus the 'performance metrics', or engineering characteristics, are determined by a complex network of design variables, or physical properties. It is in determining these basic variables that the engineering designer ultimately satisfies the customer's requirements. (Source: Ramaswamy and Ulrich, 1992.)

Example 3: car door

This example concentrates on selected attributes of one major product component, a car door. (Remember that car purchasers are said to be highly influenced in their choice of car just by the sound and feel of the door closing!) Figure 9.4 shows the first stage of developing and refining the set of product attributes from research on customer requirements. Using the objectives tree method enabled primary, secondary and tertiary levels of customer requirements

PRIMARY   SECONDARY   TERTIARY

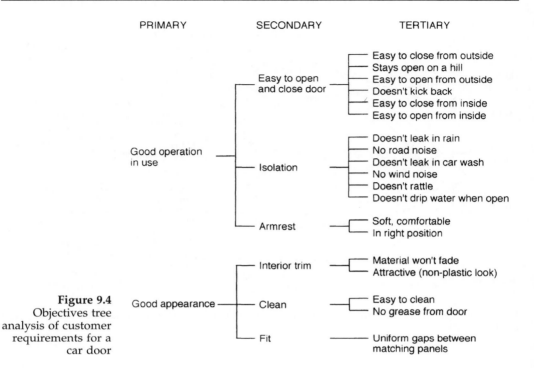

Figure 9.4
Objectives tree
analysis of customer
requirements for a
car door

to be identified and sorted into attribute 'bundles'. The relative importance weight of each attribute was also determined by market research surveys.

Using 'hall tests', customer perceptions of two competing products were established in comparison with their perceptions of the design team's own existing product. These customer perceptions were scored on a five-point scale, with a score of 5 representing the perceived best performance, and 1 representing the worst.

Part of the final, fully developed 'house of quality' is shown in Figure 9.5. The customer perceptions of the performance of competing products are shown graphically on the right.

Objective measures of the relevant engineering characteristics were determined, and are shown below the matrix for the current and two competing products. Positive and negative interactions between ECs are shown in the matrix 'roof'. Finally, on the 'bottom line' are the targets set for a redesign of the car door, after considerations not only of imputed importance but also the technical difficulty and estimated cost of making improvements on the current design. (Source: Hauser and Clausing, 1988.)

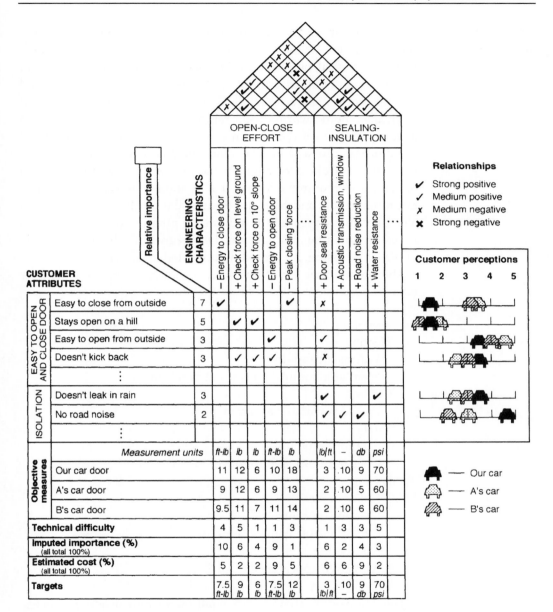

Figure 9.5 'House of quality' (partial) for a car door

Worked example: fan heater

A domestic fan heater provides a portable heat source for warming the air in domestic interiors. It is a typical product for which satisfactory technical performance has also to be matched by ease of use. Customer-valued attributes are therefore strongly influenced by engineering characteristics. Typical user requirements are that the heater should warm the air rapidly and maintain a comfortable air temperature – it therefore needs to be able to provide an initial rapid heat output and then be adjustable to a lower output, thermostatically controlled to maintain the desired temperature. Typical 'ease-of-use' requirements are that it should be easily moveable, not too big, safe and does not burn to the touch. The heater must also be visually pleasant in a domestic interior context. Since it is based around a fan, it might also be designed for the dual function of offering cooling, by providing unheated air movement.

The preliminary list of customer attributes, together with their relative importance ratings could therefore be:

Attribute	*Importance rating*
Heating	
Warms air rapidly	16
Maintains comfortable air temperature	12
Cooling	
Provides variable air movement	10
Safety	
Safe for home use	20
Does not burn skin to touch	16
Usability	
Easily moved	8
Easy-to-use controls	4
Clearly visible control settings	4
Not too big	6
Attractive appearance	4

The engineering characteristics fall under four headings: heater, fan, casing and ergonomics. Relevant characteristics of the heater element include the electrical resistance of the wire, the current and the voltage, and for the fan include fan speed, volume of air flow and air velocity provided. Relevant characteristics of the casing include the thermal insulation it provides, the design of the air outlet grille, and its form and colour. Ergonomic characteristics are related to moving and using the heater, and therefore include its overall weight, size and stability.

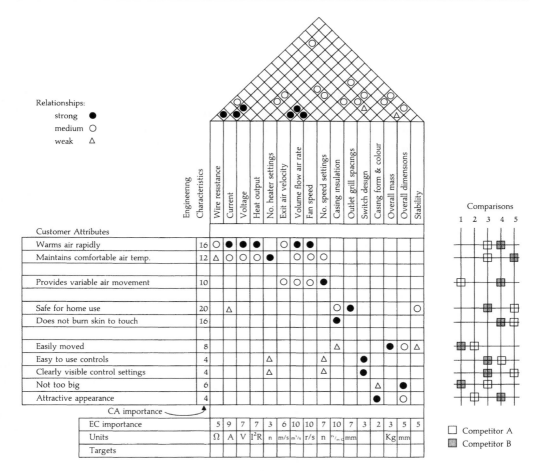

Figure 9.6 'House of quality' for a domestic fan heater

Figure 9.6 shows a 'house of quality' developed for the fan heater design. In the main body of the matrix, the strengths of relationships between CAs and ECs have been assessed as strong, medium or weak. The importance ratings for the engineering characteristics have then been determined by scoring 3 points for each strong relationship, 2 for a medium relationship and 1 for a weak relationship. These have been multiplied by the importance ratings for the CAs, totalled and normalized to indicate EC importance ratings. Although these figures have very limited mathematical validity, they do roughly indicate the relative importance of the ECs in determining the product characteristics.

In the 'roof' of the chart, interactions between ECs have also been identified, again using strong, medium or weak relationships.

The next step would be to set targets for the ECs, in the bottom row of the matrix.

In setting the targets, reference can be made to the comparisons with two competitor products, given on the right-hand side of the chart. In general, Competitor A's model is better than Competitor B's over all the CAs of safety and ease of use, but is worse on the performance CAs and on appearance. The design team's aim is therefore to design an attractive product that matches Competitor B's performance and Competitor A's ease of use and safety aspects. The relationships in the 'house of quality' between CAs and ECs indicate where the team has to concentrate its engineering design efforts in order to meet its aim.

10 *Generating Alternatives*

The generation of solutions is, of course, the essential, central aspect of designing. Whether one sees it as a mysterious act of creativity or as a logical process of problem solving, the whole purpose of design is to make a proposal for something new – something which does not yet exist.

The focus of much writing and teaching in design is therefore on novel products or machines, which often appear to have arisen spontaneously from the designer's mind. However, this overlooks the fact that most designing is actually a variation from or modification to an already existing product or machine. Clients and customers usually want improvements rather than novelties.

Making variations on established themes is therefore an important feature of design activity. It is also the way in which much creative thinking actually develops. In particular, creativity can often be seen as the reordering or recombination of existing elements.

This creative reordering is possible because even a relatively small number of basic elements or components can usually be combined in a large number of different ways. A simple example of arranging adjacent squares into patterns demonstrates this:

No. of squares	No. of distinct shape arrangements
2	1
3	2
4	5
5	12
6	35
7	108
8	369
⋮	⋮
16	13 079 255

Engineering Design Methods: Strategies for Product Design   N. Cross
© 2008 John Wiley & Sons, Ltd

The number of different arrangements, i.e. patterns or designs, soon becomes a 'combinatorial explosion' of possibilities.

The *morphological chart* method exploits this phenomenon, and encourages the designer to identify novel combinations of elements or components. The chart sets out the complete range of elements, components or subsolutions that can be combined together to make a solution. The number of possible combinations is usually very high, and includes not only existing, conventional solutions but also a wide range of variations and completely novel solutions.

The main aim of this method is to widen the search for possible new solutions. 'Morphology' means the study of shape or form; so a 'morphological analysis' is a systematic attempt to analyse the form that a product or machine might take, and a 'morphological chart' is a summary of this analysis. Different combinations of subsolutions can be selected from the chart, perhaps leading to new solutions that have not previously been identified.

The Morphological Chart Method

Procedure

List the features or functions that are essential to the product

The purpose of this list is to try to establish those essential aspects that must be incorporated in the product, or that it must be capable of doing. These are therefore usually expressed in rather abstract terms of product requirements or functions. In the morphological chart method they are sometimes called the design 'parameters'. As with many other design methods, instead of thinking in terms of the physical components that a typical product might have, you have to think of the functions that those components serve.

The items in the list should all be at the same level of generality, and they should be as independent of each other as possible. They must also comprehensively cover the necessary functions of the product or machine to be designed. However, the list must not be tool long; if it is, then the eventual range of possible combinations of subsolutions may become unmanageably large. About four to eight features or functions would make a sensible and manageable list.

For each feature or function list the means by which it might be achieved

These secondary lists are the individual subsolutions which, when combined, one from each list, form the overall design solution. These subsolutions can also be expressed in rather general terms, but it is probably better if they can be identified as actual components or physical embodiments. For instance, if one of the 'functions' of a vehicle is that it has motive power, then the different 'means' of achieving this might be engines using different fuels – e.g. petrol, diesel, electricity, gas.

The lists of means can include not only the existing, conventional components or subsolutions of the particular product, but also new ones that you think might be feasible.

Draw up a chart containing all the possible subsolutions

The morphological chart is constructed from the previous lists. At first, this is simply a grid of empty squares. Down the left-hand side are listed the essential features or functions of the product – i.e. the first list made earlier. Then across each row of the chart are entered the appropriate secondary lists of subsolutions or means of achieving the functions. There is no relationship within the columns of the chart; the separate squares are simply convenient locations for the separate items. There might be, say, three means of achieving the first function, five means of achieving the second function, two means of achieving the third function, and so on.

When it is finished, the morphological chart contains the complete range of all the theoretically possible different solution forms for the product. This complete range of solutions consists of the combinations made up by selecting one subsolution at a time from each row. The total number of combinations is therefore often very large. For instance, if there were only three rows (functions), with three squares (means) in the first row, five in the second and two in the third, then the complete set of possible combinations would number $3 \times 5 \times 2 = 30$. Because of this potential combinatorial explosion, the list of means for each function should be kept reasonably short.

Identify feasible combinations of subsolutions

Clearly, for any product the complete range of possible combinations can be a very large number. Some of these combinations – probably a small number – will be existing solutions; some will be feasible new solutions; and some – possibly a great number – will be impossible solutions, for reasons of practicality or because particular pairs of subsolutions may be incompatible.

If the total number of possible combinations is not too large, then it may be possible to list each combination, and so to set out the complete range of solutions. Each potential solution can then be considered, and one or more of the better solutions (for reasons of cost, performance, novelty, or whatever attributes are important) chosen for further development.

If, as is more likely, the total number of possible combinations is very large, then some means has to be found of reducing this to something more manageable. One way of doing this is to choose only a restricted set of subsolutions from each row – say, those that are known to be efficient or practical, or look promising for some other reason. Another way is to identify the infeasible subsolutions, or incompatible pairs of subsolutions, and so rule out those combinations that would include them.

A really exhaustive search of all the possible combinations in a morphological chart requires much patient and tedious work. The only alternative is a more intuitive, or perhaps random, search of the chart for solutions.

Summary

The aim of the morphological chart method is to generate the complete range of alternative design solutions for a product, and hence to widen the search for potential new solutions. The procedure is as follows:

1. List the features or functions that are essential to the product. Whilst not being too long, the list must comprehensively cover the functions, at an appropriate level of generalization.

2. For each feature or function list the means by which it might be achieved. These lists might include new ideas as well as known existing components or subsolutions.

3. Draw up a chart containing all the possible subsolutions. This morphological chart represents the total solution space for the product, made up of the combinations of subsolutions.

4. Identify feasible combinations of subsolutions. The total number of possible combinations may be very large, and so search strategies may have to be guided by constraints or criteria.

Examples

Example 1: potato harvesting machine

An improvement to the often rather abstract and wordy form of morphological charts can be made by using – where possible – pictorial representations of the different means for achieving the functions. An example is shown in Figure 10.1: a morphological chart for a potato harvesting machine. One selected combination of subsolutions is highlighted in the chart. Notice that two subsolutions are chosen for the subfunction 'separate stones'. This suggests

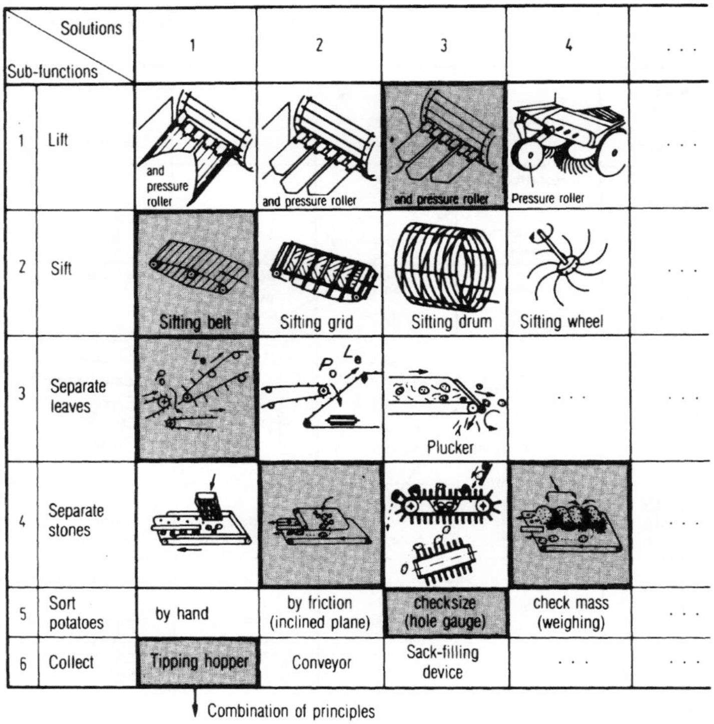

Figure 10.1
Morphological chart
for a potato
harvesting machine.
The shaded boxes
show one possible
combination of
subsolutions

either that each of these two subsolutions is not really adequate on
its own, or else that the morphological chart itself has not really
been constructed carefully enough – perhaps 'separate stones' is
not just one subfunction, but needs to be more carefully defined.
(Source: Pahl and Beitz, 1999.)

Example 2:
welding positioner

A welding positioner is a device for supporting and holding
a workpiece and locating it in a suitable position for welding.
Figure 10.2 shows a morphological chart for such a device, using
words augmented by sketch diagrams. One possible combination
of subsolutions is indicated by the zigzag line through the chart.
Even then, it was found that there were alternatives for the
actual embodiment of some of the subsolutions. For example, the
sketches in Figure 10.3 show alternative configurations for the
chosen means of enabling the tilting movement. (Source: Hubka,
1982.)

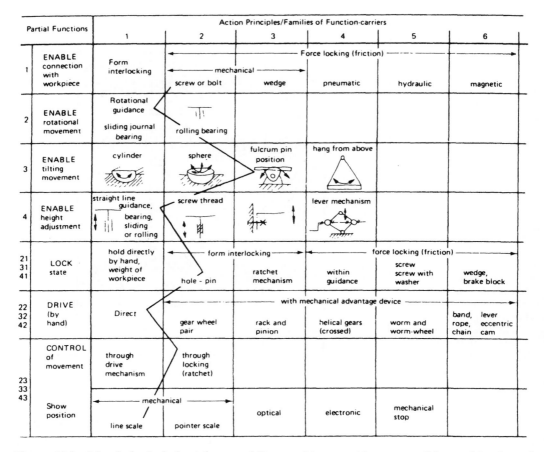

Figure 10.2 Morphological chart for a welding positioner, with one possible combination of subsolutions picked out by the zigzag line

Example 3: shaft coupling

This example shows that even small components can be usefully subject to morphological analysis. The example is that of a shaft coupling similar to the conventional 'Oldham' coupling, which transmits torque even in the case of radial and axial offsets of the shafts. Figure 10.4 shows a part of the morphological chart that was drawn up. One solution type (A) was analysed into its components and elements (presented here in columns, rather than rows) and the various subsolutions listed in pictures and words. Two new alternative combinations (B and C) are shown by the different sets of dots in the squares of the chart. One of these (B) was developed and patented as a novel design, as shown in Figure 10.5. (Source: Ehrlenspiel and John, 1987.)

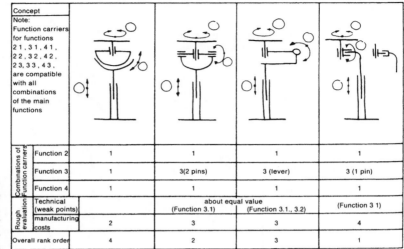

		Concept				
		Function 2	1	1	1	1
Combinations of Function carriers		Function 3	1	3(2 pins)	3 (lever)	3 (1 pin)
		Function 4	1	1	1	1
Rough evaluation		Technical (weak points)		about equal value (Function 3.1)	(Function 3.1., 3.2)	(Function 3 1)
		manufacturing costs	2	3	3	4
		Overall rank order	4	2	3	1

Figure 10.3 Four possible combinations for a welding positioner worked up into concept sketches

Note: Function carriers for functions 2 1, 3 1, 4 1, 2 2, 3 2, 4 2, 2 3, 3 3, 4 3, are compatible with all combinations of the main functions

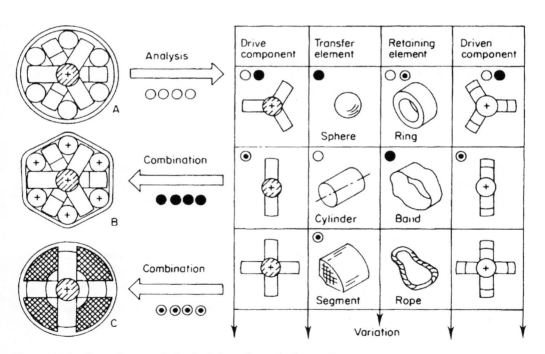

Figure 10.4   Part of a morphological chart for a shaft coupling

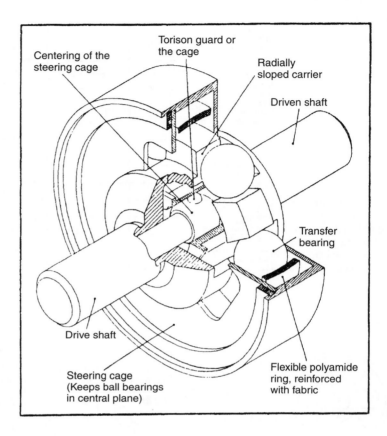

Centering of the
steering cage

Torison guard or
the cage

Radially
sloped carrier

Driven shaft

Drive shaft

Transfer
bearing

Steering cage
(Keeps ball bearings
in central plane)

Flexible polyamide
ring, reinforced
with fabric

Figure 10.5
Design for a novel
form of coupling,
derived from one of
the combinations in
the morphological
chart

*Example 4: field
maintenance
machine*

This example shows an adaptation of the principles of morphological analysis. It is concerned solely with the form arrangement,
or configuration, of the essential basic elements of the product, and
represents the alternative configurations in purely graphical terms.
The example is that of a sports field maintenance machine, and the
morphological analysis shown in Figure 10.6 explores the alternatives for configuring the elements of operator, engine, driven wheels
and nondriven wheels, and the possible disposition of these on one,
two or three vertical levels. The sketches show alternative layouts
for either a three- or four-wheel machine, with the arrangement
options systematically varied so as to generate all possible design
configurations.

The options were evaluated against design criteria, and one
preferred option developed, as shown in Figure 10.7. (Source:
Tovey, 1986.)

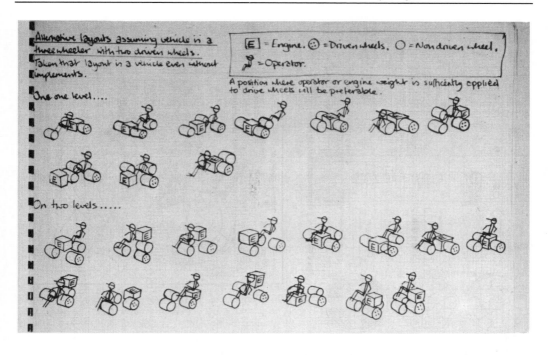

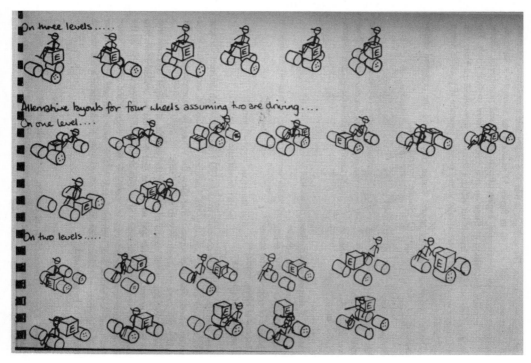

Figure 10.6   Morphological analysis of a field maintenance machine

Figure 10.7
Design sketch for
one preferred option
derived from the
morphological
analysis

*Worked example:
forklift truck*

This example is concerned with finding alternative versions of the conventional forklift truck used for lifting and carrying loads in factories, warehouses, etc. If we investigate a few of these machines we might identify the essential generic features as follows:

- Means of support which allows movement.

- Means of moving the vehicle.

- Means of steering the vehicle.

- Means of stopping the vehicle.

- Means of lifting loads.

- Location for operator.

These features seem to be common to all forklift trucks, although different versions have different means of achieving the functions. For example, most such trucks run on wheels (means of support) that allow the vehicle to go anywhere on a flat surface, but some are constrained to run on rails.

When we look at the means of moving the vehicle, we might conclude that this is too general a feature and we decide that it should be broken down into separate features for (a) the means of propulsion (normally driven wheels), (b) the power source (such as electric motor, petrol engine or diesel engine) and (c) transmission type (gears and shafts, belt, hydraulic, etc.). Adding some new and perhaps rather fanciful alternatives to the conventional alternatives would enable a list like the following to be generated:

Feature	*Means*
Support	Wheels, track, air cushion, slides, pedipulators.
Propulsion	Driven wheels, air thrust, moving cable, linear induction.
Power	Electric, petrol, diesel, bottled gas, steam.
Transmission	Gears and shafts, belts, chains, hydraulic, flexible cable.
Steering	Turning wheels, air thrust, rails.
Stopping	Brakes, reverse thrust, ratchet.
Lifting	Hydraulic ram, rack and pinion, screw, chain or rope hoist.
Operator	Seated at front, seated at rear, standing, walking, remote control.

A morphological chart incorporating these lists is shown in Figure 10.8. You might like to calculate how many possible different solution combinations there are in this chart.

Feature	Means				
Support	Wheels	Track	Air cushion	Slides	Pedipulators
Propulsion	Driven wheels	Air thrust	Moving cable	Linear induction	
Power	Electric	Petrol	Diesel	Bottled gas	Steam
Transmission	Gears and shafts	Belts	Chains	Hydraulic	Flexible cable
Steering	Turning wheels	Air thrust	Rails		
Stopping	Brakes	Reverse thrust	Ratchet		
Lifting	Hydraulic ram	Rack and pinion	Screw	Chain or rope hoist	
Operator	Seated at front	Seated at rear	Standing	Walking	Remote control

Figure 10.8
Morphological chart
for forklift trucks

There are a staggering 90 000 possible forklift truck designs in the chart. Of course, some of these are not practicable solutions, or imply incompatible options – for example, an air cushion vehicle could not have steering by wheels. A typical, conventional forklift truck would comprise the following set of options from the chart:

Support	Wheels
Propulsion	Driven wheels
Power	Diesel engine
Transmission	Gears and shafts
Steering	Turning wheels
Stopping	Brakes
Lifting	Rack and pinion
Operator	Seated at rear.

The inclusion of a few unconventional options in the chart suggests some possibilities for radical new designs. For instance, the idea of 'pedipulators' (i.e. walking mechanisms similar to legs and feet) might lead to designs suitable for use on rough ground such as building sites – or even capable of ascending flights of steps.

The chart can also be used to help generate somewhat less fanciful but nonetheless novel design ideas. For example, the idea

Figure 10.9
One selected combination of subsolutions from the morphological chart

Feature	Means				
Support	Wheels	Track	Air cushion	Slides	Pedipulators
Propulsion	Driven wheels	Air thrust	Moving cable	Linear induction	
Power	Electric	Petrol	Diesel	Bottled gas	Steam
Transmission	Gears and shafts	Belts	Chains	Hydraulic	Flexible cable
Steering	Turning wheels	Air thrust	Rails		
Stopping	Brakes	Reverse thrust	Ratchet		
Lifting	Hydraulic ram	Rack and pinion	Screw	Chain or rope hoist	
Operator	Seated at front	Seated at rear	Standing	Walking	Remote control

of using rails for steering might well be appropriate in some large warehouses, where the rails could be laid in the aisles between storage racks. The vehicle would have wheels for support and for providing propulsion. It would be electrically powered since it would be used indoors. One of the problems of electric vehicles is the limited battery power, so we might propose that our new design would pick up power from a live electric rail – like electric trains. This might be feasible in a fully automated warehouse which would not have the safety problems associated with people having to cross the rails. The 'operator' feature would therefore be remote control. A compatible set of subsolutions for this new design therefore becomes:

Support	Wheels
Propulsion	Driven wheels
Power	Electric motor
Transmission	Belt
Steering	Rails
Stopping	Brakes
Lifting	Screw
Operator	Remote control.

This set is shown as a selection from the morphological chart in Figure 10.9.

11 *Evaluating Alternatives*

When a range of alternative designs has been created, the designer is then faced with the problem of selecting the best one. At various points in the design process there may also be decisions of choice to be made between alternative subsolutions or alternative features that might be incorporated into a final design. Choosing between alternatives is therefore a common feature of design activity.

Choices can be made by guesswork, by 'intuition', by experience or by arbitrary decision. However, it is better if a choice can be made by some more rational, or at least open, procedure. Not only will the designer feel more secure in making the choice, but others involved in decision-making, such as clients, managers and colleagues in the design team, will be able to participate in, or assess the validity of, the choice.

If some of the previous design methods have already been used in the design process, then there should be some information available which should guide a choice between alternatives. For example, design proposals can be checked against criteria established by the performance specification method; and if design objectives have been established by the objectives tree method then these can be used in the evaluation of alternative designs.

In fact, the evaluation of alternatives can only be done by considering the objectives that the design is supposed to achieve. An evaluation assesses the overall 'value' or 'utility' of a particular design proposal with respect to the design objectives. However, different objectives may be regarded as having different 'values' in comparison with each other – i.e. may be regarded as being

Engineering Design Methods: Strategies for Product Design   N. Cross
© 2008 John Wiley & Sons, Ltd

more important. Therefore it usually becomes necessary to have some means of differentially weighting objectives, so that the performances of alternative designs can be assessed and compared across the whole set of objectives.

The *weighted objectives* method provides a means of assessing and comparing alternative designs, using differentially weighted objectives. This method assigns numerical weights to objectives, and numerical scores to the performances of alternative designs measured against these objectives. However, it must be emphasized that such weighting and scoring can lead the unwary into some very dubious arithmetic. Simply assigning numbers to objectives, or objects, does not mean that arithmetical operations can be applied to them. For instance, a football player assigned the number 9 is not necessarily three times as good as, or worth three times as much as a player assigned the number 3 – even though the player may score three times as many goals! Arithmetical operations can only be applied to data which have been measured on an *interval* or *ratio* scale.

The Weighted Objectives Method

Procedure

In order to make any kind of evaluation it is necessary to have a set of criteria, and these must be based on the design objectives –

List the design objectives

i.e. what it is that the design is meant to achieve. These objectives should have been established at an early point in the design process. However, at the later stages of the process – when evaluation becomes especially important – the early set of objectives may well have become modified, or may not be entirely appropriate to the designs that have actually been developed. Some clarification of the set of objectives may therefore be necessary as a preliminary stage in the evaluation procedure.

The objectives will include technical and economic factors, user requirements, safety requirements, and so on. A comprehensive list should be drawn up. Wherever possible, an objective should be stated in such a way that a quantitative assessment can be made of the performance achieved by a design on that objective. Some objectives will inevitably relate to qualitative aspects of the design; these may later be allocated 'scores', but the earlier warning about limitations on the use of arithmetic must be remembered.

Rank-order the list of objectives The list of objectives will contain a wide variety of design requirements, some of which will be considered to be more important than others. As a first step towards determining relative weights for the objectives, it is usually possible to list them in a rank order of importance. One way of doing this is to write each objective on a separate card and then to sort the cards into a comparative rank order, i.e. from 'most important' to 'least important'.

As with many other aspects of this design method, it is usually helpful if the rank ordering of objectives can be done as a team effort, since different members of a design team may well give different priorities to different objectives. Discussion of these differences will (hopefully!) lead to a team consensus. Alternatively, the client may be asked to decide the rank ordering, or market research might be able to provide customers' preferences.

The rank-ordering process can be helped by systematically comparing pairs of objectives, one against the other. A simple chart can be used to record the comparisons and to arrive at a rank order, like this:

Objectives	A	B	C	D	E	Row totals
A	—	0	0	0	1	1
B	1	—	1	1	1	4
C	1	0	—	1	1	3
D	1	0	0	—	1	2
E	0	0	0	0	—	0

Each objective is considered in turn against each of the others. A figure 1 or 0 is entered into the relevant matrix cell in the chart, depending on whether the first objective is considered more or less important than the second, and so on. For example, start with objective A and work along the chart row, asking 'Is A more important than B?'...'than C?'...'than D?', etc. If it is considered more important, a 1 is entered in the matrix cell; if it is considered less important, a 0 is entered. In the example above, objective A is considered less important than all others except objective E.

As each row is completed, so the corresponding column can also be completed with an opposite set of figures; thus, if row A reads 0 0 0 1 then column A must be 1 1 1 0. If any pair of objectives is considered equally important, a ½ can be entered in both relevant squares.

When all pairs of comparisons have been made, the row totals indicate the rank order of objectives. The highest row total indicates the highest priority objective. In the example above, the rank order therefore emerges as:

B

C

D

A

E

It is here that one of the first problems of ranking may emerge, where relationships may not turn out to be transitive. That is, objective A may be considered more important than objective B, and objective B more important than objective C, but objective C may then be considered more important than objective A. Some hard decisions may have to be made to resolve such problems!

A rank ordering is an example of an *ordinal* scale; arithmetical operations cannot be performed on an ordinal scale.

Assign relative weightings to the objectives

The next step is to assign a numerical value to each objective, representing its weight relative to the other objectives. A simple way of doing this is to consider the rank-ordered list as though the objectives are placed in positions of relative importance, or value, on a scale of, say, 1 to 10 or 1 to 100. In the example above, the rank-ordered objectives might be placed in relative positions on a scale of 1 to 10 like this:

10   B

9

8

7   C

6

5   D

4   A

3

2   E

1

The most important objective, B, has been given the value 10, and the others then given values relative to this. Thus, objective C is valued as about 70% of the value of objective B; objective A is valued twice as highly as objective E, etc. The corresponding scale values are the relative weights of the objectives. (Note that the highest and lowest ranked objectives are not necessarily placed at the absolute top and bottom positions of the scale.)

If you can achieve such relative weightings, and feel confident about the relative positions of the objectives on the scale, then you have converted the ordinal rank-order scale into an *interval* value scale, which can be used for arithmetic operations.

An alternative procedure is to decide to share a certain number of 'points' – say, 100 – amongst all the objectives, awarding points on relative value and making tradeoffs and adjustments between the points awarded to different objectives until acceptable relative allocations are achieved. This can be done on a team basis, with members of the team each asked to allocate, or 'spend', a fixed number of total points between the objectives according to how highly they value them. If 100 points were allocated amongst objectives A to E in the earlier example, the results might be:

B	35
C	25
D	18
A	15
E	7

An objectives tree can be used to provide what is probably a more reliable method of assigning weights. The highest-level, overall objective is given the value 1.0; at each lower level, the subobjectives are then given weights relative to each other but which also total 1.0. However, their 'true' weights are calculated as a fraction of the 'true' weight of the objective above them.

This is clarified in Figure 11.1. Each box in the tree is labelled with the objective's number (O_0, O_1, O_{11}, etc.), and given two values: its value relative to its neighbours at the same level, and its 'true' value or value relative to the overall objective. Thus, in this example, objectives O_{111} and O_{112} are given values relative to each other of $0.67 : 0.33$; but their 'true' values can only total 0.5 (the 'true' value of objective O_{11}) and are therefore calculated as $0.67 \times 0.5 = 0.34$ and $0.33 \times 0.5 = 0.16$.

Using this procedure it is easier to assign weights with some consistency because it is relatively easy to compare subobjectives in small groups of two or three and with respect to a single higher-level objective. All the 'true' weights add up to 1.0, and this also ensures the arithmetical validity of the weights.

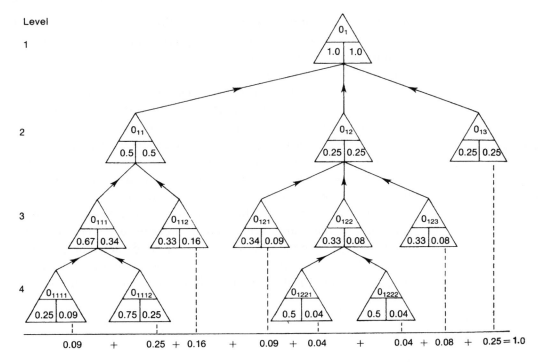

Figure 11.1 Use of an objectives tree for assigning relative weights to sub-objectives

Establish performance parameters or utility scores for each of the objectives

It is necessary to convert the statements of objectives into parameters that can be measured, or at least estimated with some confidence. Thus, for instance, an objective for a machine to have 'high reliability' might be converted into a performance parameter of 'breakdowns per 10 000 hours running time', which might be either measured from available data or estimated from previous experience with that type of machine.

Some parameters will not be measurable in simple, quantifiable ways, but it may be possible to assign utility scores estimated on a points scale. The simplest scale usually has five grades, representing performance that is

- Far below average
- Below average
- Average
- Above average
- Far above average.

Often, a 5-point scale (0–4) is too crude, and you will need to use perhaps a 9-point (0–8) or 11-point (0–10) scale. The degrees of performance assessed by an 11-point and a 5-point scale might be compared as in Table 11.1.

Both quantitative and qualitative parameters can be compared together on a points scale, representing the worst to best possible performance range. For example, the fuel consumption and, say, the comfort of a motorcar could be represented on a seven-point scale as in Table 11.2.

Table 11.1 Comparison of 11-point and 5-point evaluation scales.

11-point scale	Meaning	5-point scale	Meaning
0	Totally useless solution	0	Inadequate
1	Inadequate solution		
2	Very poor solution	1	Weak
3	Poor solution		
4	Tolerable solution		
5	Adequate solution	2	Satisfactory
6	Satisfactory solution		
7	Good solution	3	Good
8	Very good solution		
9	Excellent solution	4	Excellent
10	Perfect or ideal solution		

Table 11.2 Comparison of quantitative and qualitative parameters.

Points	Fuel consumption (miles/gallon)	Comfort
0	<25	Very uncomfortable
1	26–29	Poor comfort
2	30–33	Below average comfort
3	34–37	Average comfort
4	38–41	Above average comfort
5	42–45	Good comfort
6	>46	Very comfortable

Care must be taken in compiling such points scales, because the values ascribed to the parameters may not rise and fall linearly. For example, on the scale in Table 11.2, the value of fuel consumption is assumed to increase linearly, but it might well be regarded as more valuable to provide improvements in fuel consumption at the lower end of the scale rather than the upper end. That is, the 'utility curve' for a parameter might be an exponential or other curve, rather than linear.

Calculate and compare the relative utility values of the alternative designs

The final step in the evaluation is to consider each alternative design proposal and to calculate for each one a score for its performance on the established parameters. Once again, the participation of all members of the design team is recommended (and especially those whose views ultimately count, such as customers!), since different solutions may be scored differently by different people.

The raw performance measures or points scores on each parameter for each alternative design must be adjusted to take account of the different weights of each objective. This is done by simply multiplying the score by the weight value, giving a set of adjusted scores for each alternative design which indicates the relative 'utility value' of that alternative for each objective.

These utility values are then used as a basis of comparison between the alternative designs. One of the simplest comparisons that can be made is to add up the utility value scores for each alternative. These total scores then allow the alternatives to be ranked in order of overall performance.

Other comparisons are possible, such as drawing graphs or histograms to represent the utility value 'profiles' of the alternative designs. These visual, rather than numerical, comparisons present a 'picture' which may be easier to absorb and reflect on. They also highlight where alternatives may be significantly different from each other in their performance.

The benefit of using this evaluation method often lies in making such comparisons between alternatives, rather than using it simply to try to choose the 'best' alternative. Many rather contentious weightings, points scores and other decisions will probably have been made in compiling the evaluation, and some of the arithmetic may well be highly dubious. The 'best' overall utility value may therefore be highly misleading; but the discussions, decisions, rankings and comparisons involved in the evaluation are certain to have been illuminating.

Summary

The aim of the weighted objectives method is to compare the utility values of alternative design proposals, on the basis of performance against differentially weighted objectives. The procedure is as follows:

1. List the design objectives. These may need modification from an initial list; an objectives tree can also be a useful feature of this method.

2. Rank-order the list of objectives. Pair-wise comparisons may help to establish the rank order.

3. Assign relative weightings to the objectives. These numerical values should be on an interval scale; an alternative is to assign relative weights at different levels of an objectives tree, so that all weights sum to 1.0.

4. Establish performance parameters or utility scores for each of the objectives. Both quantitative and qualitative objectives should be reduced to performance on simple points scales.

5. Calculate and compare the relative utility values of the alternative designs. Multiply each parameter score by its weighted value: the 'best' alternative has the highest sum value. Comparison and discussion of utility value profiles may be a better design aid than simply choosing the 'best'.

Examples

Example 1: city car

In this example, a design team undertook a study of the 'city car' concept – i.e. a small runabout for use in cities or other limited-journey purposes. Many different solutions have been developed to this problem, with varying degrees of success. The design study included an analysis of the features of the many previous examples of city car designs, as well as market research, town planning and engineering criteria, etc. As a part of the design study, the designers drew up a morphological chart of six basic types of city car and the variants within each type for aspects such as the positioning of the engine. This total set of variants was then evaluated, using weighting factors and an evaluation of each variant on a scale of 0 to 10 (Figure 11.2). From this, car type 4 emerges as the preferred basic form, and was used as the concept for more detailed design work. (Source: Pighini *et al.*, 1983.)

	Sub-function	Function carrier	Task	Principle of evaluation	Weight factor	Type 1	
A	Offers space, support, protection for people and luggage	Body and frame of car	Optimal form for car	Internal space Protective ability	0 12	5	
B	Generates the power for transmission	Motor and trans-mission	Optimal position for motor	Available space Complexity of transmission	0 08	7	8
C	Supports people in a safe and comfortable way	Seats	Optimal disposition for seats	Safety Comfort Possibility of getting 4 seats	0 08	4	5
D	Offers space for luggage	Luggage room	Optimal position and higher capacity of luggage room	Space used for luggage Internal room	0 08	5	4
E₁ E₂	For entry and exit For views outside	Doors Windows	Optimal number, dimensions and position of doors and windows	Facilities for entry, exit and putting in luggage Visibility for driver and passengers	0 08	7	7
F	Changes direction of car driving	Steering system	Optimal disposition of steering system	Position of motor Complexity of steering system	0 08	9	10
G			Aesthetic evaluation		0 08	9	
H			Cost evaluation		0 16	4	
I			Safety evaluation		0 24	4	5
			Total sum			5 48	5 88
			Order of merit			9	7

NOTES 1 Evaluation marks 0 unacceptable, 1-3 still acceptable; 4-6 fair; 7-9 good, 10 very good (optimal solution)
2 Total sum (mark · weighting factor)
3 *The marks in parentheses are for the car with two seats

Figure 11.2 Evaluation chart for alternative concepts for a small city car

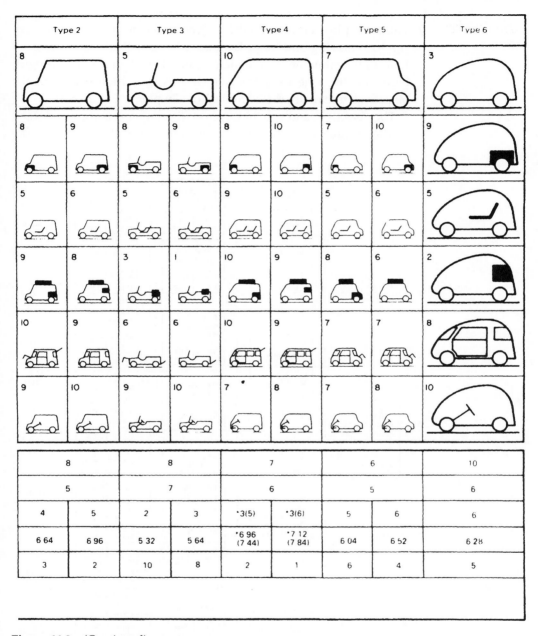

Figure 11.2   (Continued)

Example 2: test rig This example is taken from the project for the design of a laboratory rig for carrying out impulse-load tests on shaft connectors. A thorough evaluation was made of a number of alternative designs, based on the objectives tree which was presented earlier in the objectives tree method (Example 3 of Chapter 6; Figure 6.4).

The objectives and subobjectives at different levels were weighted in the manner described in the procedure (see Figure 11.3). The

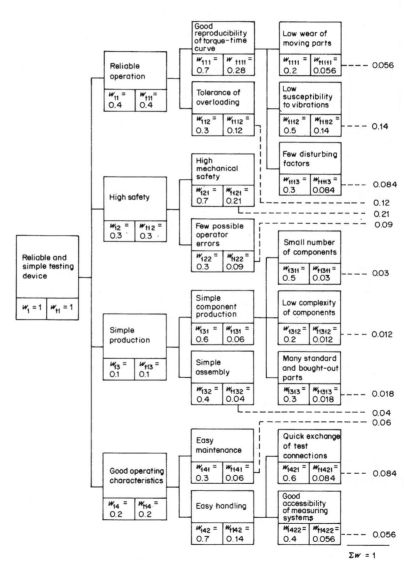

Figure 11.3
Weighted values assigned to an objectives tree for an impulse-load testing machine

design team then went on to devise measurable or assessable parameters for all of the objectives, as indicated in the comparison chart (Figure 11.4).

Utility values were calculated for each objective, for each of four alternative designs. The second alternative (variant V2) emerges as the 'best' solution, with an overall utility value of 6.816. However, variant V3 seems quite comparable, with an overall utility value of 6.446. A comparison of the 'value profiles' of these two alternatives was therefore made. This is shown in Figure 11.5, where the thickness of each bar in the chart represents the relative weight of each objective, and its length represents the score for that objective achieved by the particular design.

The chart shows that V2 has a more consistent profile than V3, with fewer relatively weak spots in its profile. V2 therefore seems to be a good 'all-round' design, and the comparison confirms it as being the best of the alternatives. However, improvement of V3 in perhaps just one or two of its lower-scoring parameters might easily push it into the lead. (Source: Pahl and Beitz, 1999.)

Example 3: van rain shield

The aim of this design project was to develop a rain protection accessory for a van, covering the area between the opened rear doors, attached to the van and opening automatically as the doors open. The context for such a device is that, during loading and unloading of a van, the operator, the cargo and the van interior get wet if it is raining. Also, some people use a van as a mobile workshop, using the tailgate area to work in. The rain shield is to provide protection for this tailgate area. There was no previous rain protection solution of this type available.

The design team researched the relevant materials, mechanisms, dimensions, fixing points and user requirements. They then generated a number of alternative concepts, and ten were developed to a formal evaluation stage. The evaluation requirements and their weighting factors, between 1 and 5, are listed in Figure 11.6.

The three concepts with the highest fulfilment of requirements are shown in Figure 11.7, and an evaluation matrix for these concepts is shown in Figure 11.8. Each concept was given a score for each of the requirements, and this was multiplied by the weight factor to give a value. In calculating the sum values, concepts A and C emerged as very close in meeting the requirements. (Source: Axeborn *et al.*, 2004.)

No.	Evaluation criteria	Wt.	Parameters	Unit	Variant V_1 Magn. m_{i1}	Value V_{i1}	Weighted value WV_{i1}	Variant V_2 Magn. m_{i2}	Value V_{i2}	Weighted value WV_{i2}	Variant V_3 Magn. m_{i3}	Value V_{i3}	Weighted value WV_{i3}	Variant V_4 Magn. m_{i4}	Value V_{i4}	Weighted value WV_{i4}
1	Low wear of moving parts	0.056	Amount of wear	-	high	3	0.168	low	6	0.336	average	4	0.224	low	6	0.336
2	Low susceptibility to vibrations	0.14	Natural frequency	s^{-1}	410	3	0.420	2370	7	0.980	2370	7	0.980	<410	2	0.280
3	Few disturbing factors	0.084	Disturbing factors	-	high	2	0.168	low	7	0.588	low	6	0.504	(average)	4	0.336
4	Tolerance of overloading	0.12	Overload reserve	%	5	5	0.600	10	7	0.840	10	7	0.840	20	8	0.960
5	High mechanical safety	0.21	Expected mechan. safety	-	average	4	0.840	high	7	1.470	high	7	1.470	very high	8	1.680
6	Few possible operator errors	0.09	Possibilities of operator errors	-	high	3	0.270	low	7	0.630	low	6	0.540	average	4	0.360
7	Small number of components	0.03	No. of components	-	average	5	0.150	average	4	0.120	average	4	0.120	low	6	0.180
8	Low complexity of components	0.012	Complexity of components	-	low	6	0.072	low	7	0.084	average	5	0.060	high	3	0.036
9	Many standard and bought-out parts	0.018	Proportion of standard and bought-out components	-	low	2	0.036	average	6	0.108	average	6	0.108	high	8	0.144
10	Simple assembly	0.04	Simplicity of assembly	-	low	3	0.120	average	5	0.200	average	5	0.200	high	7	0.280
11	Easy maintenance	0.06	Time and cost or maintenance	-	average	4	0.240	low	8	0.480	low	7	0.420	high	3	0.180
12	Quick exchange of test connections	0.084	Estimated time needed to exchange test connections	min	180	4	0.336	120	7	0.588	120	7	0.588	180	4	0.336
13	Good accessibility of measuring systems	0.056	Accessibility of measuring systems	-	good	7	0.392	good	7	0.392	good	7	0.392	average	5	0.280
		$\Sigma w = 1.0$				$OV_1 = 51$ $R_1 = 0.39$	$OWV_1 = 3.812$ $WR_1 = 0.38$		$OV_2 = 85$ $R_2 = 0.65$	$OWV_2 = 6.816$ $WR_2 = 0.68$		$OV_3 = 78$ $R_3 = 0.60$	$OWV_3 = 6.446$ $WR_3 = 0.64$		$OV_4 = 68$ $R_4 = 0.52$	$OWV_4 = 5.388$ $WR_4 = 0.54$

Figure 11.4   Completed evaluation chart for four alternative designs for an impulse-load testing machine

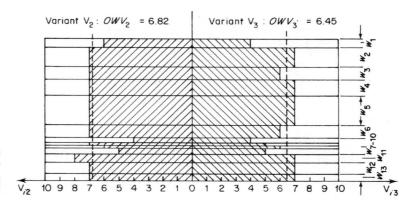

Figure 11.5
Value profiles for
alternative test rig
designs V2 and V3

Requirement	Weight factor	Description
Technology:		
Feasibility	5	Is the concept possible to realise?
Innovative	1	Does the concept give a feeling of being a brand new idea?
Function:		
Quality feeling	4	How rigid does the concept seem to be? Is it worth the
Protection	5	price?
Other functions	2	How big is the rain protection? Does it have any joints?
Possible to disengage	2	Has the protection any other function?
		Is it possible to open the doors without unfolding the rain
Possible to open one	4	protection?
door	3	Is it possible to open just one of the two doors?
Possible to open door		Is it possible to open one or both doors 180° without having
180°		to disengage the protection?
Geometric constraints:		
Small space needed	4	Does the concept affect the luggage compartment when
when unfolded		not in use?
Small reduction of	4	How much is the size of the door opening affected by the
opening		protection?
Manufacturing:		
Easy to mount	2	Is it easy to mount the protection to the van?
Easy to assemble	2	How complicated is the concept? Is it easy to assemble the
		different parts?
Low cost	5	Can existing manufacturing technologies be used, or do
		new have to be developed? Are expensive materials used?
Ergonomics:		
Easy to use	4	Does the protection unfold automatically? Is it easy to
		unfold manually? Are one or two hands needed?
Safe to use	5	Is it safe to use? Is there a risk of getting injured, e.g.
		pinched or cut?
Design:		
Not affecting van design	3	Is it possible to integrate the design of the protection into
negatively		the design of the van?
No more than 15 parts	2	How many parts does the concept include?
per side		

Figure 11.6   Evaluation requirements for a van rain shield

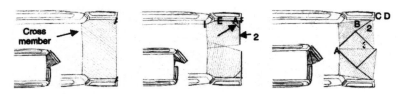

Figure 11.7   Three best concepts for the van rain shield

Requirement	Weight factor	Concept A score	Concept A value	Concept B score	Concept B value	Concept C score	Concept C value
Feasibility	5	4	20	3	15	5	25
Innovative	1	2	2	5	5	4	4
Quality feeling	4	3	12	5	20	4	16
Protection	5	5	25	3	15	2	10
Other functions	2	1	2	1	2	1	2
Possible to disengage	2	2	4	2	4	3	6
Possible to open one door	4	1	4	5	20	5	20
Possible to open door 180°	3	1	3	5	15	4	12
Small space when unfolded	4	4	16	4	16	5	20
Small reduction of opening	4	4	16	3	12	3	12
Easy to mount	2	4	8	3	6	3	6
Easy to assemble	2	5	10	2	4	2	4
Low cost	5	5	25	3	15	3	15
Easy to use	4	4	16	5	20	5	20
Safe to use	5	5	25	2	10	3	15
Not affecting van design negatively	3	3	9	5	15	5	15
No more than 15 parts per side	2	5	10	2	4	2	4
Sum			207		198		206

Figure 11.8   Evaluation matrix for the van rain shield

Example 4: bicycle splashguard   The design project for a detachable, rear-wheel splashguard for mountain bicycles, to protect riders from splashes when using the bicycle in town, was introduced in the quality function deployment (QFD) method (Example 1 of Chapter 9). Following the QFD analysis, six alternative design concepts were developed for the splashguard, including tyre-cleaning brush attachments, mud flaps of various kinds and a guard attached to saddle and seat post. In order to evaluate these alternative concepts and select the best for development, a method similar to weighted objectives was used.

This variant on the standard method uses a 'datum' design concept, against which all the others are compared. The datum may be chosen from the new alternatives under consideration, or an existing design may be used as the datum. In this example, a standard bicycle mudguard with quick-release attachments was chosen as the datum.

	Wt	I	II	III	IV	V	VI	VII
Easy attach	7	+	+	+	+	+	S	D
Easy detach	4	–	+	+	+	+	S	A
Fast attach	3	+	+	+	+	+	S	T
Fast detach	1	+	+	+	+	+	S	U
Attach when dirty	3	+	+	+	+	S	S	M
Detach when dirty	1	–	+	–	+	S	+	
Not mar	10	+	+	+	+	S	S	
Not catch water	7	–	+	–	S	S	S	
Not rattle	8	–	–	–	–	S	S	
Not wobble	7	–	–	–	S	S	S	
Not bend	4	–	–	–	S	–	S	
Long life	11	–	S	–	S	–	S	
Lightweight	7	+	S	S	–	S	S	
Not release accidentally	10	+	S	S	S	S	S	
Fits most bikes	7	+	S	S	S	S	S	
Streamlined	5	–	S	–	–	+	S	
Total +		8	8	6	7	5	1	0
Total –		8	3	7	3	2	0	0
Overall total		0	5	–1	4	3	1	0
Weighted total		1	17	–15	9	5	1	0

Figure 11.9
Evaluation chart for
alternative concepts
for a bicycle
splashguard

The set of design objectives and their weights were determined
in the QFD analysis. For each objective, each alternative design
concept was then judged as either better (+), worse (–) or the same
(s), in comparison to the datum. The decision matrix is shown in
Figure 11.9. Totals for the + signs and – signs are given for each
concept at the bottom of the matrix. A weighted overall total for
each concept is calculated by summing the positive and negative
weights of the relevant objectives.

In this example, concept 2 emerged as the clear leader. However, it
has to be remembered that this is in comparison with the datum, and
that direct comparisons between the alternative concepts themselves
should also be made. Concept 2 was therefore selected as a new
'datum' and comparisons made with concepts 4 and 5. This check
confirmed that concept 2 was the preferred alternative. (Source:
Ullman, 2002.)

Worked example: reusable syringe

This example is a medical product: a general purpose, reusable syringe for patients' self-use at home, allowing them to make regular injections of drugs. It is important that the quantity of drug to be injected is precisely controlled, and a major segment of the users of such a syringe is elderly people. Accurate and easy metering of the drug dosage are therefore important attributes of the design. Although a medical product, it is also a product for mass manufacture, and cost is also an important factor.

The list of design objectives therefore comprises the following:

- ease of handling
- ease of use
- readability of dose setting
- dose metering accuracy
- durability
- portability
- ease of manufacture.

A design team has produced seven alternative initial design concepts for the syringe, based on a variety of means and mechanisms for setting and measuring the dosage and activating the injection. The team needs to evaluate the different concepts and choose one for final development.

The design team prepares an evaluation matrix for the seven initial concepts, as shown in Figure 11.10. The evaluation criteria (design objectives) are listed down the left of the matrix, and the seven alternatives (identified by a brief descriptor) are listed along the top. One concept (D, a relatively simple solution) is chosen as the datum, or reference concept against which the others will be rated. Each alternative concept is then compared with the datum, and is rated 'better than' (+), 'equal to' (0) or 'worse than' (–), on each of the design criteria. These ratings are summed for each concept, net scores calculated and the resultant relative rankings entered at the bottom of the matrix.

Although concept A emerges as the top-ranked concept, the team recognize that some of the apparently weaker concepts may nevertheless contain good features that might be useable in a final concept, and so they consider combining ideas to produce further new design concepts. They notice that concepts D and F might be

Selection Criteria	A Master Cylinder	B Rubber Brake	C Ratchet	D (reference) Plunge Stop	E Swash Ring	F Lever Set	G Dial Screw
Ease of handling	0	0	−	0	0	−	−
Ease of use	0	−	−	0	0	+	0
Readability of settings	0	0	+	0	+	0	+
Dose metering accuracy	0	0	0	0	−	0	0
Durability	0	0	0	0	0	+	0
Ease of manufacture	+	−	−	0	0	−	0
Portability	+	+	0	0	+	0	0
Sum +'s	2	1	1	0	2	2	1
Sum 0's	5	4	3	7	4	3	5
Sum −'s	0	2	3	0	1	2	1
Net Score	2	−1	−2	0	1	0	0
Rank	1	6	7	3	2	3	3
Continue?	Yes	No	No	Combine	Yes	Combine	Revise

Figure 11.10 Initial evaluation chart for seven alternative concepts for a reusable syringe

Concepts

Selection Criteria	Weight	A (reference) Master Cylinder		DF Lever Stop		E Swash Ring		G+ Dial Screw+	
		Rating	Weighted Score	Rating	Weighted Score	Rating	Weighted Score	Rating	Weighted Score
Ease of handling	5%	3	0.15	3	0.15	4	0.2	4	0.2
Ease of use	15	3	0.45	4	0.6	4	0.6	3	0.45
Readability of settings	10	3	0.3	3	0.3	5	0.5	5	0.5
Dose metering accuracy	25	3	0.75	3	0.75	2	0.5	3	0.75
Durability	15	3	0.45	5	0.75	4	0.6	3	0.45
Ease of manufacture	20	3	0.6	3	0.6	2	0.4	2	0.4
Portability	10	3	0.3	3	0.3	3	0.3	3	0.3
Total Score			3.00		3.45		3.10		3.05
Rank			4		1		2		3
Continue?			No		Develop		No		No

Figure 11.11   Final evaluation chart for selected, combined and refined concepts for the syringe

combined to remove several of their 'worse than' ratings, and they consider that concept G could be revised to improve its handling. A new evaluation matrix is therefore prepared, for comparison of the two top-ranked original concepts, A and E, the combined concept DF and the improved concept G+ (Figure 11.11).

In this final evaluation matrix, the design objectives (selection criteria) are weighted relative to each other. Concept A is chosen as the datum, or reference concept, and given a rating score of 3 points on each criterion. The other concepts are then rated against this datum: 5 points for 'much better' and 4 points for 'better' than the datum, 3 points for 'same', 2 points for 'worse' and

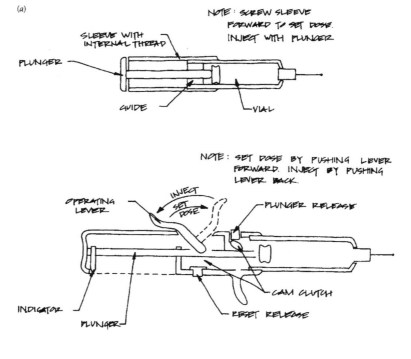

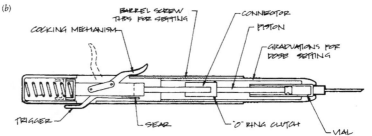

Figure 11.12
(a) Original sketches for syringe design concepts D and F. (b) Sketch of the combined concept DF selected for design development

1 point for 'much worse' than the datum. Weighted scores are calculated and summed, leading to rank positions for the four concepts, at the bottom of the matrix. The combined concept DF (shown in Figure 11.12) emerges as the preferred concept for final development. (Source: Ulrich and Eppinger, 1995.)

12 *Improving Details*

A great deal of design work in practice is concerned not with the creation of radical new design concepts but with the making of modifications to existing product designs. These modifications seek to improve a product: to improve its performance, to reduce its weight, to lower its cost, to enhance its appearance, and so on. All such modifications can usually be classified into one of two types: they are either aimed at increasing value to the purchaser or at reducing cost to the producer.

The value of a product to its purchaser is what he or she thinks the product is worth. The cost of a product to its producer is what it costs to design, manufacture and deliver it to the point of sale. A product's selling price normally falls somewhere between its cost to the producer and its value to the purchaser.

Designing is therefore essentially concerned with adding value. When raw materials are converted into a product, value is added over and above the basic costs of the materials and their processing. How much value is added depends on the perceived worth of the product to its purchaser, and that perception is substantially determined by the attributes of the product as provided by the designer.

Of course, values fluctuate, depending on social, cultural, technological and environmental contexts, which change the need for, relevance of or usefulness of a product. There are also complex psychological and sociological factors which affect the symbolic or esteem value of a product. But there are also more stable and comprehensible values associated with a product's function, and it is principally these functional values which are of concern to the engineering designer.

Engineering Design Methods: Strategies for Product Design   N. Cross
© 2008 John Wiley & Sons, Ltd

The *value engineering* method focuses on functional values, and aims to increase the difference between the cost and value of a product: by lowering cost or adding value, or both. In many cases, the emphasis is simply on reducing costs, and the design effort is concentrated onto the detailed design of components, on their materials, shapes, manufacturing methods and assembly processes. This more limited version of the method is known as *value analysis*. It is usually applied only to the refinement of an existing product, whereas the broader value engineering method is also applicable to new designs or to the substantial redesign of a product. Value analysis particularly requires detailed information on component costs.

Because of the variety and detail of information required in value analysis and value engineering, they are usually conducted as team efforts, involving members from different departments of a company, such as design, costing, marketing and production departments.

The Value Engineering Method

Procedure

List the separate components of the product, and identify the function served by each component

One of the ways in which companies seek to better their rivals' products is to buy an example of the competing product, strip it down to its individual components and try to learn how their own product might be improved in both design and manufacture. This is one way of learning some of a competitor's secrets without resorting to industrial espionage.

The same sort of technique is at the heart of value engineering and value analysis. The first analytical step in the method is to strip a product down to its separate components – either literally and physically, or by producing an itemized parts list and drawings. However, parts lists and conventional engineering drawings are of limited value in understanding and visualizing the components, the ways in which they fit together in the product overall and how they are manufactured and assembled. So if an actual product, or a prototype version, is not available for dismantling, then something like exploded diagrams of the product are helpful in showing components in three-dimensional form and in their relative locations or assembly sequences.

The purpose of this first step in the procedure is to develop a thorough familiarity with the product, its components and their assembly. This is particularly important if a team is working on the project, since different team members will have different views of the product, and perhaps only limited understandings of the components and their functions. So it is necessary to go through an exhaustive analysis of the subassemblies and individual components, and how they contribute in functional terms to the overall product.

Sometimes it is not at all clear what function a component serves or contributes to! This may be found particularly in products which have had a long life and may have gone through many different versions: some components may simply be redundant items left over from earlier versions. However, it may also be the case that components have been introduced to cope with problems that arose in the use of the product, and so any components which may appear to be redundant should not be dismissed too readily. Sometimes redundancy is even deliberately designed in to a product in order to improve its reliability.

The objective of this step in the procedure is to produce a complete list of components, grouped as necessary into subassemblies with their identified functions. In value engineering, rather than the more limited value analysis applications of this method, a similar objective applies, even though the ultimate intention might be to develop a completely new product, rather than just to make improvements to an existing product. In this case, the starting point might be an existing product against which it has been decided to compete in the market, or an 'archetypal' or hypothetically typical version of the proposed new product.

Determine the values of the identified functions

Questions of value are, of course, notoriously difficult. They are the stuff of political debate and of subjective argument between individuals. Reaching agreement in a team on the 'value' of particular product functions therefore may not be easy. However, it must be remembered that the value of a product means its value as perceived by its purchaser. So the values of product functions must be those as perceived by customers, rather than by designers or manufacturers. Market research must therefore be the basis of any reliable assessment of the values of functions.

The market prices of different products can sometimes provide indicators of the values that customers ascribe to various functions. For instance, some products exist in a range of different versions, with more functions being incorporated in the products at the higher end of the range. Differences in prices should therefore reflect differences in the perceived values of the additional functions. However, customers are likely to 'perceive' a product as a total entity, rather than as a collection of separate functions, and subjective factors such as appearance are often of more importance than objective functional factors. It is said that the solidity of the 'clunk' made by closing a car door is one of the most important factors influencing a customer's perception of the value of a motorcar.

Considerable efforts have been put into trying to quantify perceived values or benefits, particularly in connection with the 'cost–benefit analysis' method used in planning. For example, in transport planning, some of the benefits of a new road or bridge can be quantified in terms of the time saved by travellers in using the new facility. Attempts are then made to convert all such benefits (and costs) into monetary terms, so that direct comparisons can be made.

Despite the difficulty of assessing values, it is necessary to make the best attempt one can to rationalize and express the perceived values of component functions. It can be pointless to reduce the costs of components if their values are also being reduced, so that the product becomes less desirable (or 'valuable') to prospective purchasers. If quantified and reliable estimates of values cannot be made, then at least simple assessments of high/medium/low value can be attempted.

Determine the costs of the components

Surprising though it may seem, it is not always easy for a company to determine the exact costs of components used in products. The company's accounting methods may not be sufficiently specific for itemized component costs to be identifiable. One of the useful by-products of a value analysis or value engineering exercise, therefore, can be the improvement of costing methods. Team working in the exercise again becomes particularly relevant, because reliable cost information at sufficient detail may only be obtainable by synthesizing information from different departmental specialists.

It is not sufficient to know just the cost of the material in a component, or even its bought-in cost if it is obtained from a supplier. The value analysis team needs to know the cost of the component as an element of the overall product cost – i.e. after it is fully finished and assembled into the product. Therefore, as well as material or bought-in costs there are labour and machine costs to be added for the assembly processes. It is sometimes suggested that factory overhead costs should also be added, but these can be very difficult to assign accurately to individual components, and instead can perhaps be assumed to be spread equally over all components.

It is important not to ignore low-cost components, particularly if they are used in large numbers (e.g. screws or other fasteners). Even a relatively small cost reduction per item can amount to a substantial overall saving when multiplied by the number of components used.

As well as determining the absolute costs of components, their relative or percentage costs in terms of the total product cost should also be calculated. Attention might then be focused on components or subassemblies which represent a significant portion of the total cost.

Search for ways of reducing cost without reducing value, or of adding value without adding cost

This fundamental design stage calls for a combination of both critical and creative thinking. The critical thinking is aimed at what the design *is*, and the creative thinking is aimed at what it *might be*. The 'reverse engineering' concept of stripping down a competitor's product to look for ways of improving on it is a useful one to bear in mind at this stage. It is usually easier to be critical of, and to suggest improvements to someone else's design rather than one's own, and it is this kind of creative criticism that is needed at this final stage.

Attempts to reduce costs usually focus on components and on ways of simplifying their design, manufacture or assembly, but the functions performed by a product should also be looked at critically, because it may be possible to simplify them, reduce their range, or even eliminate them altogether if they are of limited value to the purchaser.

There are some general strategies which can be applied in order to direct the search for ways of reducing costs. The first is to concentrate on high-cost components, with a view to substituting lower-cost alternatives. The second is to review any components used in large numbers, since small individual savings may add up to a substantial overall saving. A third strategy is to identify components and functions which are matched as high-cost/high-value, or low-cost/low-value, since the aim is to achieve high-value functions with low-cost components. One particular technique is to compare the cost of a component used in the design with the absolute lowest-cost means of achieving the same function; large differences suggest areas for cost reduction, even though the lowest-cost version may not be a viable option.

A checklist of cost-reduction guidelines can be followed:

Eliminate	Can any function, and therefore its components, be eliminated altogether? Are any components redundant?
Reduce	Can the number of components be reduced? Can several components be combined into one?
Simplify	Is there a simpler alternative? Is there an easier assembly sequence? Is there a simpler shape?
Modify	Is there a satisfactory cheaper material? Can the method of manufacture be improved?
Standardize	Can parts be standards rather than specials? Can dimensions be standardized or modularized? Can components be duplicated?

Whilst the value analysis approach tends to emphasize reducing costs, the broader value engineering approach looks also for ways of adding value to a product. For example, rather than eliminating functions, as suggested above, value engineering might seek ways of improving or enhancing a product's functions. Nevertheless, the aim is always to increase the value/cost ratio.

One of the most significant means of adding value to a product, without necessarily increasing its cost, is to improve its ease of use. This has become particularly evident with the preference for personal computers which are found to be 'user friendly'. In this case, the 'friendliness' perhaps applies more to the computer's software than its hardware, or at least to the combination of software and hardware such that use of the computer seems natural and easy. However, similar principles can be applied to all machines – their use should be straightforward, clear and comfortable. There is a considerable body of knowledge in the field of ergonomics which can be applied to these user aspects of machine design.

Other attributes which commonly contribute to the quality or value of a product are:

Utility	Performance on aspects such as capacity, power, speed, accuracy or versatility.
Reliability	Freedom from breakdown or malfunction; performance under varying environmental conditions.

Safety   Secure, hazard-free operation.

Maintenance   Simple, infrequent or no maintenance requirements.

Lifetime   Except for disposable products, a long lifetime which offers good value for the initial purchase price.

Pollution   Little or no unpleasant or unwanted by-products, including noise and heat.

Finally, there is a whole class of value attributes related to aesthetics. This includes not only the appearance of a product – colour, form, style, etc. – but also aspects such as surface finish and feel to the touch.

Evaluate alternatives and select improvements

The application of value analysis or value engineering should result in a number of alternative suggestions for changes to the product design. Some of these alternatives might well be incompatible with each other, and in fact all suggestions should be carefully evaluated before selecting those which can be shown to be genuine improvements.

Summary

The aim of the value engineering method is to increase or maintain the value of a product to its purchaser whilst reducing the cost to its producer. The procedure is as follows:

1. List the separate components of the product, and identify the function served by each component. If possible, the actual product should be disassembled into its components; exploded diagrams and component function charts are more useful than parts lists.

2. Determine the values of the identified functions. These must be the values as perceived by customers.

3. Determine the costs of the components. These must be after fully finished and assembled.

4. Search for ways of reducing cost without reducing value, or of adding value without adding cost. A creative criticism is necessary, aimed at increasing the value/cost ratio.

5. Evaluate alternatives and select improvements.

Examples

Example 1: ceiling diffuser

Substantial cost savings can often be made even on relatively simple products. Although the cost per unit may not be great, the total savings can be large when a large number of units is involved. In this example, the product is a $10 ceiling diffuser, the device which covers the ceiling output points of heating and air conditioning systems. Its function is to help spread the air flow into the room, and it should look attractive.

Informal discussions with a number of users and customers (including installers), to determine their likes and dislikes about the existing diffuser, revealed several areas where the designers felt that there were mismatches between manufacturing costs and perceived value to the customers of certain details. The designers' recommendations for changes are given in Figure 12.1.

The changes resulted in reducing material costs by 24% and labour costs by 84%, saving the company nearly $500 000. Functional and aesthetic improvements also resulted from the redesign. These, plus a 20% price reduction, helped achieve a significant increase in the company's market share. (Source: Fowler, 1990.)

Function	Cost	Design Change
Style product	$1.34 (13%)	Delete the center cone.
Minimize housekeeping	$0.05 (0.5%)	Users complained about ceiling smudging caused by air output. The team reshaped the remaining three cones to feather the air gradually, to prevent it from contacting the ceiling around the diffuser.
Ensure stability	$1.36 (13%)	Team referred to its function-cost worksheet and found that half of the webs, web clips and rivets, all of the springs, and much of the assembly labour was to ensure stability. Changed to two wireforms with two of the legs spot-welded to a newly designed center cone.
Ease installation and simplify adjustment	$1.07 (10%)	Both of these functions are performed by access areas in the cones, to permit the use of a screwdriver to attach the unit to the ductwork and to adjust the ductwork damper after installation. Installation access areas were deleted, since modern installation does not require screw attachment. A hole was added to permit damper adjustment.
Protect shipments	$0.86 (8%)	The prestudy discussions with users/ customers revealed that diffusers were invariably ordered in pairs. The shipping carton was redesigned to carry two units at a significant cost reduction.

Figure 12.1   Value analysis of an air conditioning ceiling diffuser

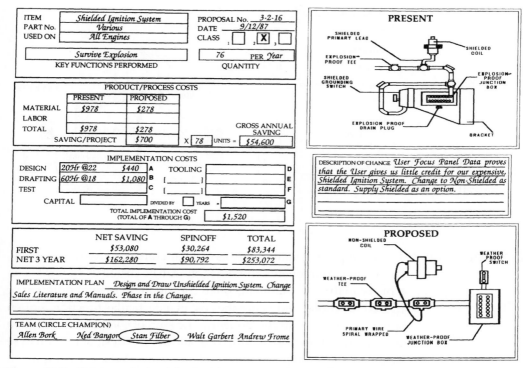

Figure 12.2   Value analysis of the ignition system of an engine

Example 2: ignition system

The product under investigation in this example was a 1200-horse power engine, used primarily in compressor systems. The engine incorporated several features or subsystems, not provided by competitors. One of these was a totally shielded ignition system, to prevent any possibility of a spark igniting any flammable gas around the engine. The company assumed that this was a valued feature of its engine. However, questionnaires to a panel of users, as part of a value engineering exercise, revealed that the shielded ignition system was not rated highly as a feature by the users.

Figure 12.2 shows the value analysis report drawn up by the design team. They proposed changing to an unshielded ignition system, with the shielded system offered as a higher-cost option for those customers who wanted it. The estimated cost of making the change was only $1520 for design and drafting. Gross annual saving was $54 000, with a $253 000 net saving over the first three years. (Source: Fowler, 1990.)

Parts or assemblies	Cost (£)
Banjo assembly	1.07
Valve body	6.62
Spring	0.39
Diaphragm assembly	2.14
Cover	2.24
Lug	0.10
Nuts, bolts and washers	2.34
Assembly cost	4.58
Total	£19.48

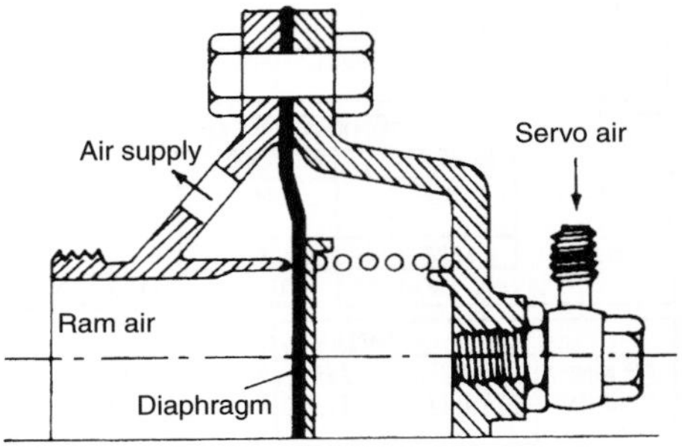

Figure 12.3
Cost analysis of an
aircraft air valve

Example 3: air valve   This example, the value analysis of an aircraft air valve (Figure 12.3), shows how components and functions can be costed in a comparison table or matrix. Components often contribute to several different or related functions, and hence the cost of a particular function is often spread across several components. The kind of component/function cost matrix shown in Figure 12.4 allows the designer to analyse in detail these often complex relationships. When a component contributes to more than one function, it may be difficult to break down its overall cost into precise part-costs per function. Approximate but well-informed estimates then have to be made.

The analysis in this example revealed the relatively high cost of the 'connect parts' function, as well as the redundancy of some elements. A redesign enabled some substantial reductions to be made, with a total cost saving of over 60 %. (Source: EITB.)

Example 4: piston   The elimination of unnecessary parts can be a significant factor in reducing the overall cost of an assembly, and is a principal focus of value analysis. Figure 12.5 shows the redesign of a small piston assembly, eliminating or combining several parts that were in the original. Separate fasteners should be eliminated wherever possible, and it was found in this example that the two screws could be eliminated by changing the cover plate from steel to plastic with a snap-fit onto the main block. The cover was also redesigned to incorporate the piston stop in a one-piece item. In the redesign, the number of parts was thus almost halved, resulting in reduced material and assembly costs with no loss of performance and an aesthetically improved product. (Source: Redford, 1983.)

(a)

Parts	Stop air	Sense ram air	Sense servo air	Sense cabin air	Connect parts	Provide mounting	Resist corrosion	Provide support	Provide interchangeability	No function	£ total cost	%
Banjo assembly		0 2		0.4					0 47		1 07	5 5
Valve body	0 4	1 0			2 82	0 8	0 2	0 8		0 6	6 62	34 0
Spring										0 39	0 39	2 0
Diaphragm assembly	0 6	0 1	0 1	0 1	0 94		0 2	0.1			2 14	11 0
Cover			0 4		1.2	0 1	0 1	0 34	0 1		2 24	11 5
Lug										0.1	0 1	0.5
Nuts, bolts and washers					2 14		0 1		0 1		2 34	12 0
Assembly					4 58						4 58	23 5
Total	1 0	1 1	0 7	0 1	12 08	0 9	0 6	1 24	0.67	1 09	19 48	100 0
% total	5 1	5 7	3 4	0 5	62 0	4 6	3 1	6 4	3 4	5 6		
High or low					H				H			

(b)

Parts	Stop air	Sense ram air	Sense servo air	Sense cabin air	Connect parts	Provide mounting	Resist corrosion	Provide seal	Provide interchangeability	Provide testing	£ total cost	%
Cover and connection	0 15	0 25	0 50	0 10	0 25	0 30		0 15	0 06		1 76	25 5
Body assembly	0 15	0 20	0 25	0 45	0 45	0 40		0 25	0 03		2 18	31 5
Diaphragm assembly	0 15	0 10	0 25	0 20	0 25	0.10		0 20	0 03		1 28	18 5
Valve assembly	0.05		0 05	0 05	0 15			0 31	0 05		0 66	9 5
Fasteners, nut bolts, etc					1 04						1 04	15 0
Total	0 50	0 55	1 05	0 80	2 14	0 80		0 91	0 17		6 92	100 0
% total	7 2	7 9	15 1	11 6	30 9	11 6		13 2	2 5			
High or low					H							

Figure 12.4 (a) Function/cost analysis matrix for the air valve. (b) Function/cost analysis matrix for the redesigned air valve

Example 5: tubular heater In a company manufacturing various kinds of electrical appliances, its range of tubular heaters was selected for a value engineering exercise. These heaters are simple and robust and used mainly in industrial and office premises to provide background heat. The product range consists of similar tubes of various lengths providing various heat outputs at a standard wattage per unit length.

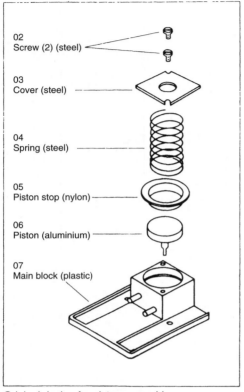

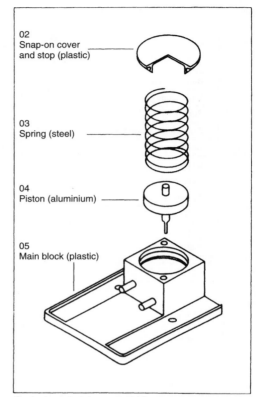

Original design for piston assembly   Redesigned piston assembly

Figure 12.5   Redesign of a piston assembly to reduce the number of components

A component/function/cost analysis, shown in Figure 12.6, revealed that the largest parts and labour cost was accounted for by what was regarded as the third most important function – that of providing the power connection. A closer examination of this function revealed two distinct subfunctions: firstly, providing an interconnector to allow tubes to be banked together on one mains connection; and secondly, providing a complex terminal connection.

An ideas-generating session produced suggestions for redesign which are shown in Figures 12.7 and 12.8. The moulded interconnector was replaced with three separate wires and a cover piece (Figure 12.7), which also enabled the terminal itself to be considerably simplified (Figure 12.8). Together, the modifications resulted in a cost reduction of 21 %.

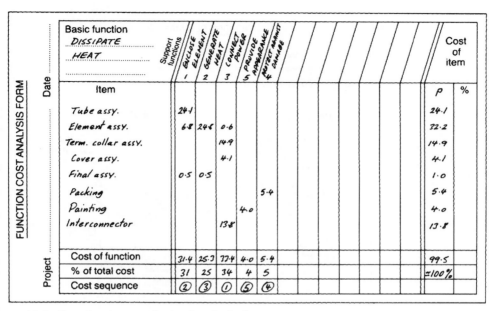

The table in the figure (FUNCTION COST ANALYSIS FORM):

	Basic function DISSIPATE HEAT	Support functions	EXCLOSE ELEMENT 1	GENERATE HEAT 2	CONNECT POWER 3	PROVIDE APPEARANCE 4	PROTECT AGAINST DAMAGE 5									Cost of item P	%
Item																	
Tube assy.			24·1													24·1	
Element assy.			6·8	24·8	0·6											32·2	
Term. collar assy.				14·9												14·9	
Cover assy.				4·1												4·1	
Final assy.			0·5	0·5												1·0	
Packing						5·4										5·4	
Painting					4·0											4·0	
Interconnector				13·8												13·8	
Cost of function			31·4	25·7	17·4	4·0	5·4									99·5	
% of total cost			31	25	34	4	5									≈100%	
Cost sequence			②	③	①	⑤	④										

(Left margin: FUNCTION COST ANALYSIS FORM, Date, Project)

Figure 12.6 Function/cost analysis of a tubular heater

Worked example: handtorch

This relatively simple example demonstrates the principles both of applying value analysis with the objective of reducing a product's cost and of applying value engineering with the objective of generating a more highly-valued, innovative product.

Figure 12.9 shows how both value analysis and value engineering projects might start: with an exploded diagram of the product, which in this case is a conventional handtorch. The diagram shows the separate components and indicates how they are assembled together in the complete product.

Market research showed that two main aspects of a torch are highly valued by users. These are, firstly, the quality of the emitted light, perceived by users as being influenced by (apart from battery power) the bulb and the reflector, and secondly the ease of use of the torch, determined by the torch body and the switch. One low-valued feature of this particular torch design was the hanging loop on the base of the torch, which was hardly used at all and thus thought by most users to be redundant.

The components, their functions and perceived values are listed in Table 12.1, with values categorized simply as high, medium or low. It is useful to note that some components which may be important to the technical performance of the product are not necessarily perceived

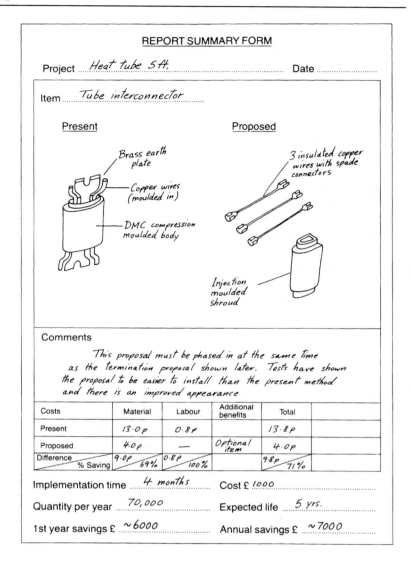

Figure 12.7
Proposals for
redesign of a tube
interconnector

The figure content:

REPORT SUMMARY FORM

Project ....Heat tube 5ft....   Date ....

Item ....Tube interconnector....

| Present | Proposed |

Brass earth plate
Copper wires (moulded in)
DMC compression moulded body

3 insulated copper wires with spade connectors
Injection moulded shroud

Comments

This proposal must be phased in at the same time as the termination proposal shown later. Tests have shown the proposal to be easier to install than the present method and there is an improved appearance

Costs	Material	Labour	Additional benefits	Total	
Present	13·0 p	0·8 p		13·8 p	
Proposed	4·0 p	—	Optional item	4·0 p	
Difference % Saving	9·0 p 69%	0·8 p 100%		9·8 p 71%	

Implementation time ....4 months....   Cost £ ....1000....

Quantity per year ....70,000....   Expected life ....5 yrs....

1st year savings £ ....~6000....   Annual savings £ ....~7000....

as of high value by users – examples here include the bulbholder and the pressure spring in the base.

A value analysis exercise led fairly quickly to some suggested modifications which would lower the product's cost without lowering its value. The reflector cover seemed to be too complicated, with its three separate components: glass, washer and screw-on retainer. A one-piece, clear plastic cap was suggested as an alternative. The base of the torch also seemed to be a rather complicated assembly, and again a one-piece plastic screw-on cap was suggested, with an integral plastic tongue spring to provide the pressure on the batteries, and the

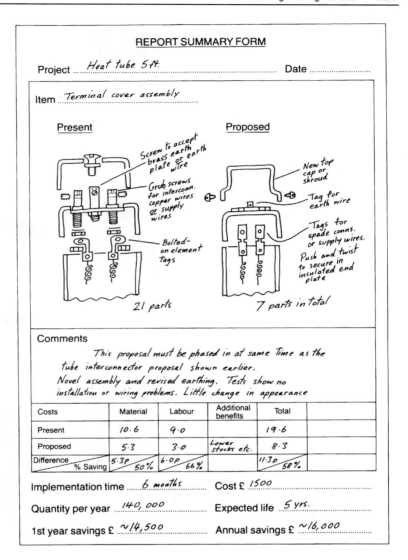

REPORT SUMMARY FORM

Project ...... *Heat tube 5 ft.* .... Date ....

Item ...... *Terminal cover assembly* ....

Present   Proposed

Screw to accept brass earth plate or earth wire

Grub screws for interconn. copper wires or supply wires

Bolted-on element Tags

New top cap or shroud

Tag for earth wire

Tags for spade conns. or supply wires.

Push and twist to secure in insulated end plate

21 parts   *7 parts in total*

Comments

    This proposal must be phased in at same time as the tube interconnector proposal shown earlier.
Novel assembly and revised earthing. Tests show no installation or wiring problems. Little change in appearance

Costs	Material	Labour	Additional benefits	Total	
Present	10.6	9.0		19.6	
Proposed	5.3	3.0	*Lower stocks etc.*	8.3	
Difference / % Saving	5.3p / 50%	6.0p / 66%		11.3p / 58%	

Implementation time ...... *6 months* ......   Cost £ *1500* ....

Quantity per year ...... *140,000* ......   Expected life ...... *5 yrs.* ......

1st year savings £ ...... *~14,500* ......   Annual savings £ *~16,000* ......

Figure 12.8
Proposals for redesign of a terminal cover assembly

hanging loop eliminated. A proposal was also made to eliminate the switch, with electrical interrupt being provided instead by twisting the head of the torch. However, on evaluation it was decided that this was not very convenient for the user, and risked losing the highly valued ease of use of the thumb-switch.

Table 12.1 shows that the costed redesign indicated a potential saving in manufacturing costs of approximately 20 %.

A more comprehensive value engineering exercise would have concentrated on the high-value aspects of the torch as perceived by users, and would have sought to improve these features, to enhance

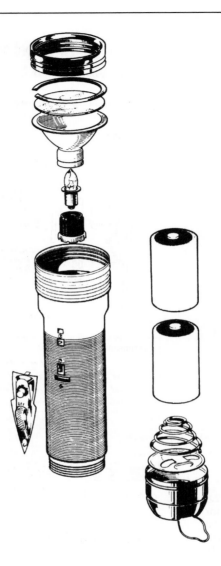

Figure 12.9
Exploded diagram of
a handtorch

them or to generate innovations related to them. The high-value features of the torch are to do with its light beam and its handling.

Some research with users might well have found that the conventional torch has some shortcomings in these areas. For instance, it seems basically designed to throw a moderately wide beam over a fairly large distance – such as for illuminating a footpath. But most use of a torch these days is for closer illumination such as finding a keyhole or making emergency repairs on a motorcar engine. In the latter type of case, it is important to be able to place the torch

Table 12.1 Components, functions and perceived values for a handtorch redesign.

Component	Function	Value	Cost (£)	
			Original	Redesign
Cap, washer, glass	Protect bulb and reflector	Medium	0.16	0.08
Reflector	Project light beam	High	0.12	0.12
Bulb	Provide light	High	0.10	0.10
Bulb holder	Hold bulb, provide electrical contact	Low	0.05	0.05
Torch body	Contain batteries, locate parts, provide hand grip	High	0.26	0.26
Switch	Provide electrical interrupt	High	0.08	0.08
Spring washer	Provide pressure on batteries	Low	0.10	0.10
Cap	Protect batteries	Medium	0.10	
Loop	Provide for hanging	Low	0.03	—
Total			1.00	0.79

down, leaving one's hands free, and to direct the beam to the appropriate spot. The conventional, cylindrical torch is poorly designed for this; it is also inconveniently shaped for carrying in a pocket or handbag.

The novel Durabeam torch (Figure 12.10) illustrates how these principles might have been applied in the design of a new product. The batteries are placed side-by-side instead of end-to-end, creating a flat, rectangular, compact body shape. The thumb-switch has been eliminated by using a 'flip-top' mechanism which acts as a switch and which also allows the angle of direction of the beam to be adjusted.

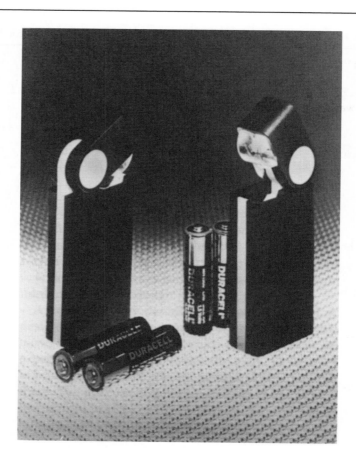

Figure 12.10
Durabeam
handtorch

Part Three
Managing Design

13 *Design Strategies*

What is a Design Strategy?

Using any particular design method during the design process will often appear to be diverting effort from the central task of designing. However, this is exactly the importance of using such a method: it involves applying some thought to the *way* in which the problem is being tackled. It requires some strategic thinking about managing the design process.

A design strategy describes the general plan of action for a design project and the sequence of particular activities (i.e. the tactics, or design methods) which the designer or design team expect to undertake to carry through the plan. To have a strategy is to be aware of where you are going and how you intend to get there. The purpose of having a strategy is to ensure that activities remain realistic with respect to the constraints of time, resources, etc., within which the design team has to work.

Many designers seem to operate with no explicit design strategy. However, having no apparent plan of action can be a strategy, of sorts! It might be called a 'random search' strategy, and might very well be appropriate in novel design situations of great uncertainty, where the widest possible search for solutions is being made. Examples of such novel situations might be trying to find applications for a completely new material, or designing a completely new machine such as a domestic robot.

For these kinds of situations, an appropriate strategy would be to search (at least to begin with) as widely as possible, hoping to find or generate some really novel and good ideas. The relevant tactics would be drawn mainly from the creative methods.

Engineering Design Methods: Strategies for Product Design N. Cross
© 2008 John Wiley & Sons, Ltd

At the opposite extreme to 'random search' would be a completely predictable or 'prefabricated' sequence of tried-and-tested actions. Such a strategy would be appropriate in familiar and well-known situations. Again, it might not seem to be an explicit strategy, simply because it involves following a well-worn path of conventional activities. Examples of appropriate situations for such a strategy might include designing another variation of the machine that the designer's employer always makes, or designing a specific and conventional type of product for an identified sector of the market.

In such situations, the design strategy would be aimed at narrowing the search for solutions and quickly homing in on a satisfactory design. Relevant tactics would be drawn from conventional techniques and the rational methods.

Strategy styles   The 'random search' and 'prefabricated' strategies represent two extreme forms. In practice, most design projects require a strategy that lies somewhere between the two extremes, and contains elements of both.

The random search strategy represents a predominantly *divergent* design approach; the prefabricated strategy represents a predominantly *convergent* approach. Normally, the overall aim of a design strategy will be to converge onto a final, evaluated and detailed design proposal. But within the process of reaching that final design there will be times when it will be appropriate and necessary to diverge, to widen the search or to seek new ideas and starting points.

The overall design process is therefore convergent, but it will contain periods of deliberate divergence (Figure 13.1).

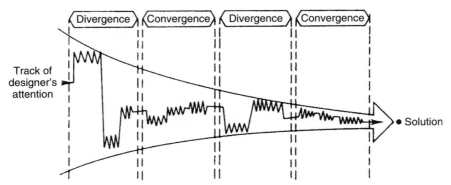

Figure 13.1   The overall design process is convergent, but it includes periods of both convergence and divergence

Psychologists have suggested that some people are more naturally convergent thinkers and some are more naturally divergent thinkers. These preferred 'thinking styles' mean that some designers may be happier with one kind of strategy style rather than with another; one person may prefer a more convergent style, whereas another may prefer a more divergent style. Alternatively, in a team context, designers with one preferred style may come to the fore in certain stages of the design process, and others come to the fore at other stages.

Convergent thinkers are usually good at detail design, at evaluation and at selecting the most appropriate or feasible proposal from a range of options. Divergent thinkers are usually good at concept design and at the generation of a wide range of alternatives. Clearly both kinds of thinking are necessary for successful design. Unfortunately, much engineering (and other) education tends to promote and develop only convergent thinking.

As well as convergent and divergent, other kinds of 'thinking styles' have also been identified by psychologists, and may also have importance in design and in the structuring of design strategies. One of the most important dichotomies in thinking style appears to be that between *serialist* and *holist*. A serialist thinker prefers to proceed in small, logical steps, tries to get every point clear or decision made before moving on to the next, and pursues a straight path through the task, trying to avoid any digressions. A holistic thinker prefers to proceed on a much broader front, picking up and using bits of information that are not necessarily connected logically, and often doing things 'out of sequence'.

Another distinction that has been made between styles of thinking is that between *linear* and *lateral* thinking. Linear thinking proceeds quickly and efficiently towards a perceived goal, but may result in getting 'stuck in a rut', whilst lateral thinking entails a readiness to see, and to move to, new directions of thought.

The dichotomies of thinking style suggested by the psychologists tend to fall into two groups:

Convergent	Divergent
Serialist	Holist
Linear	Lateral

There is even some evidence to suggest that there is a fundamental dichotomy between the 'thinking styles' of the two hemispheres of the human brain. The left hemisphere predominates in rational, verbal, analytic modes of thought, whilst the right hemisphere predominates in intuitive, nonverbal, synthetic modes of thought.

Differences of thinking styles therefore appear to be an inherent characteristic of human beings. Most people *tend* towards a preference for one style rather than another; but no-one is *exclusively* limited to just one style. In particular, it is actually important to be able to change from one style to another in the course of a design project.

However, many models of the design process, such as those discussed in Chapter 3, do tend to present design as a linear, serialistic process. This may be off-putting, and even counterproductive to those designers whose own preferred thinking style tends more towards the lateral and holistic. What is needed is a more flexible, strategic approach to designing, which identifies and fosters the right kind of thinking at the right time, and within the context of the particular design project.

Strategy analogies

To convey this more flexible approach to design strategies and tactics, some authors have resorted to the use of analogies. For example, Jones (1992) has suggested that a designer is like an explorer searching for buried treasure:

> *A new problem is like an unknown land, of unknown extent, in which the explorer searches by making a network of journeys. He has to invent this network, either before he starts or as he proceeds. Design methods are like navigational tools, used to plot the course of a journey and maintain control over where he goes. Designing, like navigation, would be straightforward if one did not have to depend on inadequate information in the first place. Unlike the explorer's, the designer's landscape is unstable and imaginary, it changes form according to the assumptions he makes. The designer has to make as much sense as he can of every fragmentary clue, so that he can arrive at the treasure without spending a lifetime on the search. Unless he is very unlucky, or very stupid, he will come across the treasure long before he has searched every inch of the ground.*

Koberg and Bagnall (2003) have suggested that the designer is like a traveller, and that 'the design process is a problem-solving journey':

> *A general rule is to find and use those methods which best fit the problem as well as the abilities of the problem solver. It's a task similar to that of selecting the route, side roads and overnight stops for an auto*

trip. Just as any competent trip planner would examine the alternative routes on a map, and read through several brochures, books or articles before choosing a route for his trip, so should the problem solver review the methods available, and not be afraid to adapt any of them to his special needs.

Instead of exploring or travelling, I prefer an analogy based on football. A design team, like a football team, has to have a strategy. The football team's strategy for defeating the opposition will consist of an agreed plan to use a variety of plays or moves (i.e. techniques or methods), to be applied as the situation demands. During the game, the choice of a move, and whether or not it is successful, will depend on the specific circumstances, on the skill of the players and on the response of the opposition.

The repertoire of moves used in a game is partly decided in advance, partly improvised on the field and also amended at the half-time briefing by the team coach. The coach's role is important because the coach maintains a wider view of the game than the players can actually out there on the field. In designing, it is necessary to adopt a similar role from time to time, in reviewing the project's strategy and progress.

For you as an individual designer, or member of a design team, tackling your problem and reaching your goal will involve both the strategic skills of the coach and the tactical skills of the player. Also, like the team, you will have to make on-field and half-time reviews of your strategy to ensure that your problem does not defeat you! A design strategy, therefore, should provide you with two things:

1. A *framework* of intended actions within which to operate.

2. A management *control* function enabling you to adapt your actions as you learn more about the problem and its 'responses' to your actions.

Frameworks for Action

One 'framework', complete with appropriate methods identified and located within it, has already been suggested. That was the procedural model of the design process that is outlined in Chapter 4:

Stage in the design process	*Appropriate method*
1. Identifying opportunities	User scenarios.
2. Clarifying objectives	Objectives tree.
3. Establishing functions	Function analysis.
4. Setting requirements	Performance specification.
5. Determining characteristics	Quality function deployment.
6. Generating alternatives	Morphological chart.
7. Evaluating alternatives	Weighted objectives.
8. Improving details	Value engineering.

If it seemed to be appropriate to the specific project in hand, then you could adopt this as a complete prefabricated strategy. It comprises an eight-stage framework covering the design process from user requirements through to detail design, and a suitable tactic (a design method) for each stage. You could, of course, add, or substitute, methods in each stage. For example, you could use brainstorming instead of a morphological chart as a way of generating alternative solutions; you could use the conventional 'design-by-drawing' method instead of, or perhaps as well as, value engineering or analysis at the stage of detail design.

However, this particular framework does imply that the design process is going to be a fairly straightforward, step-by-step process. It implies a linear design process. A design strategy that is more suited to a lateral approach might be something like this:

Stage	*Tactics to be used*
1. Divergent problem exploration	Morphological chart. Brainstorming.
2. Structuring of problem	Objectives tree. Performance specification.
3. Convergence on solution	Synectics.

Another framework might be adopted from the general pattern of the creative process, as outlined in Chapter 4. This might be developed as follows:

Stage	Tactics to be used
1. Recognition	User scenarios. Brainstorming.
2. Preparation	Objectives tree. Information search. Function analysis.
3. Incubation	Taking a holiday. Talking the problem over with colleagues and friends. Tackling another problem. Enlarging the search space: counter-planning.
4. Illumination	Morphological chart. Brainstorming. Enlarging the search space: random input.
5. Verification	Performance specification. Weighted objectives.

So you see that there can be many different strategy frameworks, and many different tactical combinations of methods and techniques.

Strategy Control

The second important aspect of a successful design strategy is that it has a strong element of management control built into it. If you are working alone on a project, then this means self-management, of course. If you are working in a team then either the team leader or the whole team, collectively, must from time to time review progress and amend the strategy and tactics if necessary.

Whatever general framework is adopted for the project, it is necessary to have some further strategy control in order to avoid unnecessary time-wasting, going down blind alleys and the like. Some simple rules of strategy control are as follows.

Keep your objectives clear

In designing, it is impossible to have one set of completely fixed objectives, because ends and means are inextricably interwoven in the product you are designing. A creative resolution of a design problem often involves changing some of the original objectives. However, this does not mean that it is impossible to have any clear objectives at all. On the contrary, it is important to have your objectives clear at any time (probably in the form of an 'objectives tree'), but also to recognize that they can change as your project evolves.

Keep your strategy under review   Remember that your overall aim is to solve the design problem in a creative and appropriate way: it is not doggedly to follow a path you have set for yourself that might be leading nowhere! A design strategy needs to be flexible, adaptable and intelligent; so review it regularly. If you feel that your actions are not being very productive, or that you are getting stuck, then pause to ask yourself if there is not a better way of proceeding. Have confidence in adapting the 'tactics', the methods and techniques, to your own ways of working and to the aims and progress of the project.

Involve other people   Different people 'see' a problem in different ways, and it is often true that 'two heads are better than one'. If you are getting stuck, one of the easiest ways to sort out what is going wrong is to explain the project to someone else – a colleague or a friend. Other people, of course, are also able to offer ideas and different viewpoints on the problem which may well suggest ways to change your approach.

Keep separate files for different aspects   There will, almost certainly, be times when you have to work on several different aspects of a project in parallel. So keep separate files which allow you to switch rapidly from one aspect to another, or to take in a new piece of information in one area without distracting your work on another. One very useful file to keep is 'Solution Ideas'. You will probably come across, or have ideas for solutions at all times throughout the project, but you will need just to keep them filed until you are ready to turn your whole attention to solution concepts or details.

Setting Strategies and Choosing Tactics

The following short exercises are intended to give some practice in devising strategic frameworks and selecting appropriate tactical methods or techniques. Each exercise need only take 5–10 minutes.

Exercise 1   Your company manufactures industrial doors of various kinds. With the increased availability of electronic devices, remote controls, and so on, the company has decided to produce a new range of automatically operated doors. You have been asked to propose a set of prototype designs that will establish the basic features of this new range. Outline your design strategy and tactics.

Exercise 2   Your company manufactures packing machinery. One of the company's most valued customers is about to change its product range and will therefore need to replace its packing machinery. You will be responsible for designing this new machinery. Outline your design strategy and tactics.

Exercise 3   You have just been appointed design consultant to a company manufacturing office equipment. Its sales have fallen drastically because its designs have failed to keep up with modern office equipment trends. To re-establish its position the company wants a completely new product that will be a step ahead of all its rivals. You have to suggest what the new product should be, and produce some preliminary design proposals for a board meeting in two weeks' time. Outline your design strategy and tactics.

Discussion of exercises

Industrial doors

The change from manual to automatic doors implies that there could be scope for rethinking the company's current range, perhaps to include some door types that were not previously included. It is therefore worth putting some divergent search effort into the early stages of the design project. It is also important not to overlook the features of existing doors that are valued by customers, so quality function deployment could be used to identify the critical characteristics. My suggested strategy would be as follows.

Framework	*Tactics*
1. Problem exploration	Brainstorming. Synectics.
2. Problem specification	Quality function deployment.
3. Alternative solutions	Morphological chart.
4. Selection of alternatives	Weighted objectives.

Packing machinery This appears to be a straightforward case of redesigning an established product. There is no apparent need for radically new design concepts, so fairly conventional methods can be used. My suggested strategy would be as follows.

Framework	*Tactics*
1. Customer requirements/ problem specification	Performance specification.
2. Alternative solutions/ evaluation of alternatives	Value engineering.
3. Detail design	Conventional design-by-drawing.

Office equipment This problem does suggest the need for some radical design thinking, and pretty quickly! Creativity techniques would therefore feature strongly in the strategy. I would start with some quick focus group research to develop user scenarios, and function analysis based on some existing products. Following some generation of ideas, I think that I might try to use a modified version of the objectives tree method in a final, more convergent stage, and to work this into my presentation to the board so as to relate the choice of alternatives to the company's objectives. My suggested strategy would be as follows.

Framework	*Tactics*
1. Divergent search	User scenarios. Function analysis.
2. Alternative solutions	Brainstorming. Morphological chart.
3. Convergent selection	Objectives tree.

I hope that these brief examples give an indication of how to adopt a strategic approach to product design, using a variety of methods as the tactics of designing. The important points to remember are to devise a strategy that responds to the particular problem and situation, to keep your strategy flexible and to review its effectiveness from time to time during the design project.

14 Product Development

Product Design

The examples of product designs that have been included in this book have ranged from one-off engineering structures, such as test rigs and packaging machines, through major mass-production machines, such as cars and washing machines, to more modest consumer products, such as torches and bicycle splashguards. The range of examples, and the scope of the book, therefore reflects a concept of 'product design' that traditionally has been divided between the two (often conflicting) camps of engineering design and industrial design.

Conflicts have sometimes arisen between these two camps because of misconceptions about each other's roles. Engineering designers sometimes see industrial designers as mere 'stylists' who add the external casing and the pretty colours to the machines that they have engineered; and industrial designers sometimes see engineering designers as the suppliers of crude mechanisms that they then have to try to convert into usable and attractive products. Of course, different products require a different mix of skills; for some products the engineering element may be relatively small, whereas for others it is very large. This mix of engineering and industrial design contributions in different products is often shown diagrammatically as in Figure 14.1.

However, the increasing competition in consumer product markets and the growing awareness of the importance of design for the market has led to reinforcement of the view that successful product design can only be accomplished by an integration of the skills of both engineering and industrial designers. Products that are

Engineering Design Methods: Strategies for Product Design   N. Cross
© 2008 John Wiley & Sons, Ltd

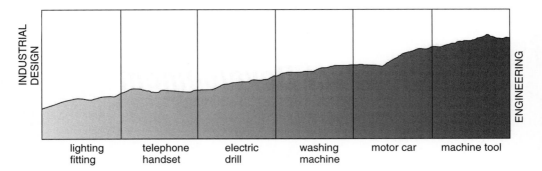

Figure 14.1   Traditional view of varying contributions of engineering and industrial design in different kinds of products

well 'engineered' but are difficult to use, ugly and unsafe are not well-designed products. Neither are products that are attractively 'styled' but are unreliable, flimsy and difficult to maintain and reuse. Good product design practice is therefore converging towards the 'industrial design engineer', a designer (or design team) with knowledge and skills from both engineering and industrial design.

Product Planning

Whether a particular product is seen as an exercise predominantly in engineering design or in industrial design, it is important to realize that 'design' is only a part of a larger process of product planning and development. This larger process extends from the business strategy of the client company, through the manufacturing, marketing and distribution of the product. Ultimately, the commercial success of a product depends upon the purchasing decision of the consumer.

From the consumers' point of view, there is little input that they can make to product planning and design. The important decisions are taken by others, and all that the consumer can do is to exercise some choice between products at the point of sale. Collectively, however, consumers are extremely important to manufacturers. The mass of consumers, all making their individual purchase decisions, constitute 'the market' for which the manufacturers plan their products. They engage in market research of various kinds and they respond to the pressures exerted by their more successful competitors in the market. The manufacturers must constantly review and develop their business strategies, of which one of

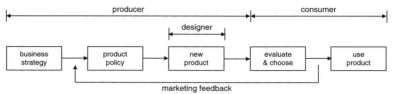

Figure 14.2
Roles of producer,
consumer and
designer

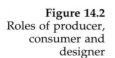

the most important elements is their strategy for introducing new products.

The situation is shown diagrammatically in Figure 14.2. The manufacturer's or producer's role starts from the establishment of a business strategy, from which new product policies and plans are developed, and specific products are then designed and manufactured. The consumer's (rather limited) role is to choose and use the available products. The consumer's decisions are, however, influential on future new product development, through the feedback loop of market research.

Clearly, this diagram is an inadequate representation of the full set of activities in product planning, specification, design, development and manufacture. The 'new product' box bridges a huge region in the middle between the producer's product planning options and the consumer's purchase choice. Major sections of the whole new product development process lie in this region.

The producer's product plan will only identify broad ranges or types of products. For instance, a hi-fi manufacturer's product plan might identify only the suggestion of producing a 'source component' to add to its product range, such as a CD player or MP3 player; a vacuum cleaner manufacturer's product plan might identify only a need to replace their mid-range model within a certain time period.

From these broad plans, it is necessary to generate some specific product ideas. These ideas might range from improvements to, or redesigns of, existing products, through suggestions for new additions to a product range, to completely new, innovative products. The ideas might come from the company management, design teams, marketing departments or even directly from consumers. Given the variety of ideas and potential new products, a screening and selection process will be necessary, to reduce the variety down to a smaller number of ideas which can be subjected to feasibility analyses. Eventually, a new product proposal emerges in the form of a product design brief and specification.

Given the brief for the new product proposal, the activities of product design proper can start. The key role of design in the overall process of product development is shown in Figure 14.3, from British Standard BS 7000 'Guide to Managing Product Design'.

The design brief is the link between the initial phase of identifying the need or motivation for the product and the creation phase of design, development and production. The design stage itself leads to a definition of a particular, specific product, which is then subject to further development and refinement before going into full-scale production. This development stage will consist of planning for the production and marketing of the product, as well as refinements in the detailed engineering of the product, its materials specification, and so on.

Design, production and marketing used to be thought of as rather separate activities. The designers would despatch their drawings to the production engineers, who would decide how to make the product (perhaps making some major changes in order to simplify the production process) and then pass the prototypes onto the marketing personnel, who would decide how to sell the product. It is now recognized that design, production and marketing development must proceed in parallel and with mutual interaction, if satisfactory and successful products are to reach the market on time.

The stages of design and development in Figure 14.3, therefore, should result in a product definition which includes not only the refined, detailed product design, but also plans for the production and marketing of the product. These plans will be implemented in the later production and distribution stages.

Product design therefore plays its essential role within a much broader process of new product creation. Product designing starts with the design brief and ends with the refined product definition. (In fact, of course, it may also at times be extended both into earlier phases of product planning and into later phases of product realization.) The overall process of new product creation can be regarded as comprising three phases. First there is a phase of product planning in which product policy is formulated and ideas sifted. This is followed by the product design phase (including plans for manufacture and marketing). Finally there is the product realization phase, including manufacture, distribution and sales.

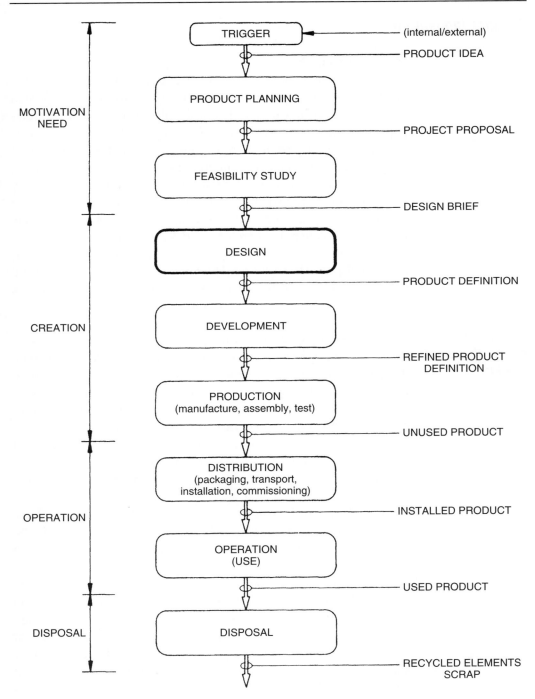

Figure 14.3 Product development process according to BS 7000

Product Innovation

Most companies have a continuous programme of product development, within which they seek to maintain and increase their product sales by continually introducing improved and updated versions of their products. But occasionally, radically new products appear on the market – products of a completely new type or form. Examples of such product innovations include ballpoint and fibre-tip pens, pocket calculators, motor scooters, personal computers, video cassette players, video cameras, etc. As well as these many successes, there are also frequent failures – product innovations that do not become economic successes, and that are soon forgotten. Product innovation is therefore a risky business, and it is not surprising that most companies prefer to stick to the safer ground of gradual product evolution. However, the rewards of successful innovation can be substantial, and so many companies are attracted, or find it necessary to venture into it.

A success story: the Sony Walkman

A classic example of successful product innovation is the Sony Walkman, the personal stereo audio cassette (and later CD) player. After the first version was introduced by Sony in 1979 there soon became hundreds of different versions of this product type, made by dozens of different companies. It seems difficult to believe now, but before the Sony Walkman there was no comparable product, and the millions of people that came to own one simply did not know that they wanted such a product before they saw it! The Walkman prefigured more recent, computer-based personal audio devices such as the Apple iPod.

The Walkman did not just appear 'out of the blue', but it did have an unconventional development history within Sony. In the 1970s, Sony was already a major manufacturer of electronic products such as radios, televisions and video and audio recorders. The Walkman concept originated with one of the company's cofounders, Masaru Ibuka, who used to listen to stereo music whilst travelling on aeroplanes, using headphones with an earlier, and much heavier, Sony portable tape recorder. In 1977 Sony launched the small, light Pressman, a portable, pocket-sized, monaural cassette tape recorder aimed specifically, as its name suggests, at journalists. Ibuka asked the engineers if they could provide him with a simple, playback-only stereo version of the Pressman. He was impressed by the resulting prototype, and showed it to his other cofounder and company chairman, Akio Morita.

After trying it, Morita was also impressed by the idea of a stereo cassette player light enough to walk around with. He instructed the Sony tape recorder division to use the Pressman as a basis for developing a new portable system. It would be a risky decision to try to sell to the public a new style of 'recording' machine, that could only play back and not record, but Morita saw the potential. One of the common features of radical product innovation is that there is an influential 'product champion' within the company, who pushes for the product against its critics, and in Sony there was no-one more influential than Morita.

Ibuka then provided another critical input. The headphones he had been using were actually much larger than the Pressman itself, and somewhat limited its 'portability'. In the Sony research laboratory, he found an ongoing, nearly completed, development project on very lightweight, miniature headphones. The crucial second step was to combine the new stereo cassette player with these lightweight miniature headphones.

At first, Sony had no idea just how successful the Walkman concept would be. They certainly did not know that Walkman would become a generic product name, like Biro for ballpoint pens or Hoover for vacuum cleaners. Only in Japan was it launched as the Walkman; in the USA it was called the Soundabout, in Scandinavia the Freestyle and in Britain the Stowaway. Only as it became such a major success did the rather strange Japanese–English of Walkman become established as the product's world-wide name. Sony had discovered, or created, a whole new market, and without doing any market research!

A failure story:
the Sinclair C5

Most product innovations that fail simply fade into obscurity and are forgotten. In contrast, the Sinclair C5 became notorious as an example of failure, probably largely because its inventor–designer, Clive Sinclair, was well known, with a reputation for radical, and usually successful, innovations in electronic products such as miniature TVs and personal computers.

The C5 was a different type of product: an electrically assisted tricycle (Figure 14.4). Sinclair had been interested in electric vehicles before, and had done some previous design work on a range of electric vehicles, including a two-seater called the C10. The stimulus for the C5 arose from some changes in UK road traffic regulations. These defined a new class of two- or three-wheeled electrically assisted pedal vehicles, which could be used on the roads by anyone

Figure 14.4
Sinclair C5
electrically assisted
tricycle. Reproduced
from Richard
Hearne

over the age of 14, without licence or insurance. This market 'niche' would include, for example, electrically assisted bicycles, but the C5 was a radically different concept. It was designed by teams of experienced engineers (for instance, the chassis and transmission were developed by Lotus Cars), and was progressive in both technology (e.g. the polypropylene body shell was the largest mass-produced injection-moulded assembly at that time) and ergonomics (e.g. the low seating position, with handlebar below the knees).

The C5 was launched in January 1985 with considerable publicity and promotion. The new company, Sinclair Vehicles, expected to sell 100 000 units per year, but production was discontinued in August 1985 with only about 5000 sold. Sinclair Vehicles went into receivership in October 1985, having lost £8.6 million.

Despite the clever engineering and design of the C5 it was a disastrous commercial failure. Clearly the concept was wrong, and people simply would not be persuaded that driving/pedalling it amongst other traffic would make them feel 'secure, but exhilarated' as the advertising claimed! Although some marketing research was done for the C5, this was *after* the essential concept had already been decided, and appeared to be mainly to aid promotion.

Technology push and market pull

Many radical product innovations seem to be based on new technology. For example, pocket calculators, personal computers and many other new electronics-based products were made possible by the development of the microprocessor chip. But, as we have seen in the success and failure stories, people's willingness to buy new products is the ultimate deciding factor – if people do not want the product then it fails. There are also many examples of new product developments that do not depend on new technology but on recognizing what people want or 'need', whether that is recyclable packaging, stacking hi-fi systems, dish washers, etc. There are therefore two strong aspects to new product development: the 'push' that comes from new technology and the 'pull' of market needs.

These two aspects are usually called *technology push* and *market pull*. Technology itself, of course, does not do any pushing. That comes from the developers and suppliers of the new technology, and from the makers of the new products. In practice, a lot of new product development is influenced by a combination of both technology push and market pull.

Many companies prefer to work on the market pull model, using market research to identify customers' wants and needs. The technology push view, on the other hand, emphasizes that innovations can create new demands and open up new markets. Market research usually cannot identify demands for products that do not yet exist.

This has been recognized particularly by those companies that try to plan new product development in terms of both technological 'seeds' and customer 'needs'; success depends on matching seeds with needs. However, even when a market need and a technology seed can be matched, and a new product concept identified, there is no guarantee that a product will actually be developed. It may require far too much financial investment, for example, or a 'product champion' may not emerge or be successful within the company. Another reason is that some product concepts are actually suppressed by companies and organizations that have a strong vested interest in maintaining the markets for their existing products. This is particularly true of industries with a heavy capital investment in the continued production of a particular product type. The motor industry, for example, failed to support the development of alternative vehicles, such as electric cars, until it began to see such innovations as potentially important to its survival.

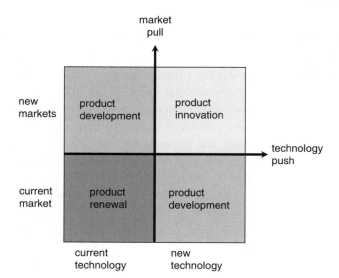

Figure 14.5
Matrix of
opportunity areas
for product planning

Some opportunities for new product development lie in the region where an already developed technology can meet an undeveloped market, while others lie in the region where new technology can be applied in an already developed market (Figure 14.5). A third region, for the most radical (and risky) product innovations, is where new technology and new market opportunities might be developed together (the upper-right sector of Figure 14.5). The Sony Walkman and Sinclair C5 were both examples of the latter.

Sony has also pursued the product development sectors, for example in modifying some of its existing products for new markets in children's goods, when it developed its My First Sony products (top-left sector of the diagram). And it has continually developed (in the bottom-right sector) new versions of the Walkman concept for its market of teenagers and young adults, by taking up new technologies such as CD and MP3 players.

I have labelled the bottom-left sector of the diagram 'product renewal'. It is where a company's current products lie. In fact, most product design activity occurs within this segment, simply because most companies cannot afford to stand still. There is a constant need for product renewal, through both minor and major product modifications, if a company is to maintain its market position. Companies fail if they cannot renew their products to match the performance of those of their competitors.

References

Archer, L.B. (1984) Systematic method for designers. In: N. Cross (ed.), *Developments in Design Methodology*. John Wiley & Sons, Ltd, Chichester.

Axeborn, M., Gould, A. and Ottosson, S. (2004) Dynamic product development of rain protection for vans. *Journal of Engineering Design*, **15**, 229–247.

Cagan, J. and Vogel, C. (2002) *Creating Breakthrough Products*. Prentice Hall, Upper Saddle River, NJ.

Cross, N. (2007) *Designerly Ways of Knowing*. Birkhäuser, Basel.

Darke, J. (1984) The primary generator and the design process. In: N. Cross (ed.), *Developments in Design Methodology*. John Wiley & Sons, Ltd, Chichester.

Davies, R. (1985) A psychological enquiry into the origination and implementation of ideas. MSc thesis, Department of Management Sciences, University of Manchester Institute of Science and Technology.

Dwarakanath, S. and Blessing, L. (1996) Ingredients of the design process. In: N. Cross, H. Christiaans and K. Dorst (eds), *Analysing Design Activity*. John Wiley & Sons, Ltd, Chichester.

Ehrlenspiel, K. and John, T. (1987) Inventing by design methodology. *International Conference on Engineering Design*. American Society of Mechanical Engineers, New York.

Engineering Industry Training Board (no date) *Value Engineering*. EITB, London.

Fricke, G. (1996) Successful individual approaches in engineering design. *Research in Engineering Design*, **8**, 151–165.

Fowler, T.C. (1990) *Value Analysis in Design*. Van Nostrand Reinhold, New York.

French, M.J. (1999) *Conceptual Design for Engineers*. Springer, London.

Grange, K. (1999) Personal communication.

Hauser, J.R. and Clausing, D. (1988) The house of quality. *Harvard Business Review*, May/June.

Hawkes, B. and Abinett, R. (1984) *The Engineering Design Process*. Pitman, London.

Hubka, V. (1982) *Principles of Engineering Design*. Butterworth, London.

Hurst, K. (1999) *Engineering Design Principles*. Arnold, London.

Jones, J.C. (1984) A method of systematic design. In: N. Cross (ed.), *Developments in Design Methodology*. John Wiley & Sons, Ltd, Chichester.

Jones, J.C. (1992) *Design Methods*. John Wiley & Sons, Inc., New York.

Koberg, D. and Bagnall, J. (2003) *The Universal Traveler*. Crisp Learning, Ontario.

Krick, E. (1976) *An Introduction to Engineering*. John Wiley & Sons, Inc., New York.

Lawson, B. (1984) Cognitive strategies in architectural design. In: N. Cross (ed.), *Developments in Design Methodology*. John Wiley & Sons, Ltd, Chichester.

Lawson, B. (1994) *Design in Mind*. Butterworth-Heinemann, Oxford.

Love, S.F. (1986) *Planning and Creating Successful Engineered Designs*. Advanced Professional Development, Los Angeles, CA.

Luckman, J. (1984) An approach to the management of design. In: N. Cross (ed.), *Developments in Design Methodology*. John Wiley & Sons, Ltd, Chichester.

March, L.J. (1984) The logic of design. In: N. Cross (ed.), *Developments in Design Methodology*. John Wiley & Sons, Ltd, Chichester.

Myerson, J. (2001) *IDEO: Masters of Innovation*. teNues, New York.

Otto, K. and Wood, K. (2001) *Product Design*. Prentice Hall, Upper Saddle River, NJ.

Pahl, G. and W. Beitz (1999) *Engineering Design*. Springer, London.

Pighini, U., di Francesco, G., Yuan, D.Z., Schettino, A.V. and Rivalta, A. (1983) The determination of optimal dimensions for a city car using methodical design with prior technology analysis. *Design Studies*, **4**(4), 233–243.

Pugh, S. (1991) *Total Design*. Addison Wesley, Wokingham.

Radcliffe, D. and Lee, T.Y. (1989) Design methods used by undergraduate engineering students. *Design Studies*, **10**(4), 199–207.

Ramaswamy, R. and Ulrich, K. (1992) Augmenting the house of quality with engineering models. In: D.L. Taylor and L.A. Stauffer (eds), *Design Theory and Methodology*. American Society of Mechanical Engineers, New York.

Redford, A.H. (1983) Design for assembly. *Design Studies*, **4**(3), 170–176.

Schön, D.A. (1983) *The Reflective Practitioner*. Basic Books, New York.

Tjalve, E. (1979) *A Short Course in Industrial Design*. Newnes-Butterworths, London.

Tovey, M. (1986) Thinking styles and modelling systems. *Design Studies*, **7**(1), 20–30.

Ullman, D.G. (2002) *The Mechanical Design Process*. McGraw-Hill, New York.

Ulrich, K.T. and Eppinger, S.D. (1995) *Product Design and Development*. McGraw-Hill, New York.

Index